SpringerBriefs in Environmental Science

More information about this series at http://www.springer.com/series/8868

T.G. Sitharam • Sreevalsa Kolathayar

Preparing for Earthquakes: Lessons for India

Springer

T.G. Sitharam
Department of Civil Engineering
Indian Institute of Science Bangalore
Bangalore, Karnataka, India

Sreevalsa Kolathayar
Civil Engineering
Amrita Vishwa Vidyapeetham University
Coimbatore, Tamil Nadu, India

ISSN 2191-5547 ISSN 2191-5555 (electronic)
SpringerBriefs in Environmental Science
ISBN 978-3-319-59521-4 ISBN 978-3-319-59522-1 (eBook)
DOI 10.1007/978-3-319-59522-1

Library of Congress Control Number: 2017942800

Printed on acid-free paper

This Springer imprint is published by Springer Nature
The registered company is Springer International Publishing AG
The registered company address is: Gewerbestrasse 11, 6330 Cham, Switzerland

Preface

Earthquakes have resulted in large-scale destruction and loss of lives worldwide. The scenario is worse in many developing countries due to poor construction practices, lack of appropriate guidelines in existing seismic codes, and poor land use planning. Most of the casualties due to earthquakes happen not because of earthquakes, but because of the poorly built structures which cannot withstand the earthquake forces. In India too, a significant portion of damage can be attributed to the lack of awareness, lack of preparedness, lack of dissemination of knowledge on the earthquake-resistant construction techniques, and high population density.

Vision

- To safeguard the lives of people during earthquakes
- To minimize the loss of assets due to earthquakes

Someday an earthquake may strike the place you live in. It cannot be predicted when it would happen, whether it will be in your lifetime or during your childrens' lifetime. When that day comes, the community must be prepared. There are various steps which individuals and communities residing in seismically active regions can take that will reduce the havoc and loss during a quake. These steps can be complementary measures, alongside other activities that focus on the progress of the community, with different stakeholders working in unison. Seismic safety must be integrated not only into construction practices, but also into community preparedness, emergency response, recovery, and long-term risk reduction efforts. However, earthquake response planning is yet to become a part of the public consciousness even in the vulnerable regions of India.

In general, the aftermaths of a disaster bring solidarity among the members of the community. There is a need to bring this spirit of togetherness and solidarity well before a disaster so that a community can be well prepared to face the disaster and thus reduce the loss of life and properties.

The past earthquakes in India have made it evident that along with structural improvement, there is a need to build resilience as well, so that society is able to take up the impact of an earthquake and recuperate fast. This involves capacity building and building a community that is aware of the earthquake hazards and is prepared for them. The central and state governments, city corporations, and residents should maintain high disaster awareness, and the inhabitants of vulnerable regions should be able to help one another, forming a community where suitable action can be taken. An earthquake disaster management system needs to be established where associated institutions can coordinate for preparedness well before the earthquakes and protect lives during an event. Stable progress of the processes to build cities that are resilient against earthquakes should be undertaken, such as the improvement of densely populated regions with wooden houses, making structures earthquake resistant, and securing lifelines, road networks, and other infrastructures.

The earthquake risk reduction and management plan should be comprehensive in nature, involving geologists to identify hazards, agencies to retrofit certain vulnerable buildings, and emergency response agencies to improve and exercise response plans. It is advised to be cautious of seismically induced landslides in hilly areas which need special consideration for either stabilization of slopes or relocation of important structures. A comprehensive plan will provide a broad picture so that various organizations can act both independently and in coordination with each other before and during the hazard. Being prepared can lessen panic, worry, and losses that come with disasters. Communities, families, and individuals must know what to do before, during, and after an earthquake event. This book sheds light on the action plans to be taken by the common public and public agencies before, during, and after earthquakes to reduce the disaster risk.

Mission

- To invoke alertness and solidarity among the public in earthquake prone areas
- To empower the community to prepare themselves to face and manage the aftermath of an earthquake

This book articulates various factors related to earthquake preparedness and develops guidelines and useful tips for communicating them to all the stakeholders. The book has been divided into four parts: (1) Background which explains earthquakes in general and the seismicity of India (2) Earthquake preparedness for individuals, families, and various stakeholders, (3) Earthquake resilient communities, where various strategies for communities to prepare themselves for a future earthquake are described and (4) Earthquake Readiness Index Tool for India, which can be applied to evaluate and enhance the earthquake preparedness level of indivdual and community. To ensure better engagement with the general public, utmost care is taken to present the report in a simple and accessible manner.

The authors thank Dr. Muralee Thummarukudy, the Chief of Disaster Risk Reduction in the UN Environment Programme for his valuable inputs on restoration strategy after an earthquake. The authors express their sincere thanks to GeoHazards International for giving permission to reproduce their contents on community resilience towards Tsunami, in context of earthquake preparedness.

Special thanks to Mr. Adheesh Damodaran for contributing the digital drawings to the manucript. Authors express their gratitutde to Prof. M L Sharma President

Indian Society of Earthquake Technology (ISET), Dr. Ajay Chourasia Secretary ISET and the Executive Committee members of ISET for their valuable comments which helped improve the content of this book. Authors acknowledge the contribution of Ms. Anupa S Anil and Mr. E Lalith Prakash towards developing Earthquake Readiness Index Tool for India which is described in Chapter 4 of this book.

Bangalore, Karnataka, India T.G. Sitharam
Coimbatore, Tamil Nadu, India Sreevalsa Kolathayar

Contents

Chapter 1
Earthquakes: The Indian Context

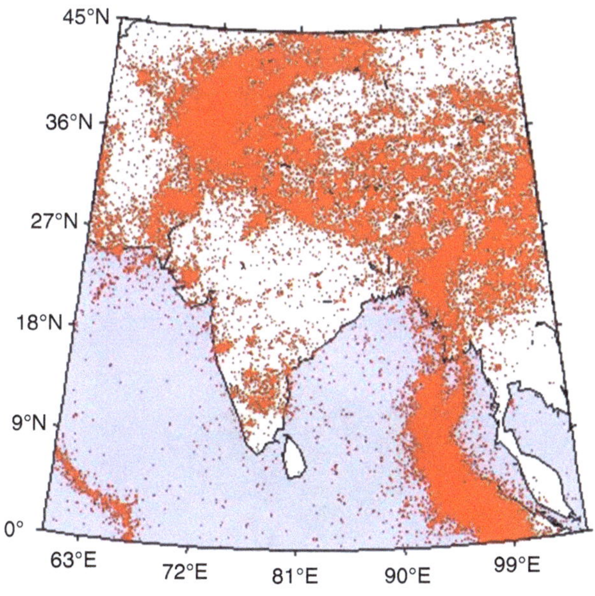

Earthquakes are unpredictable and deadly. Let us work together to save lives & property.

1.1 Introduction

Nature has its own ways of expressing itself and natural phenomena have always been a wonder to mankind. Many natural processes are looked upon as natural disasters mainly because human beings have not been able to respond to them appropriately. Earthquake is one such disaster that has affected the mankind the most. Due to

T.G. Sitharam, S. Kolathayar, *Preparing for Earthquakes: Lessons for India*,
SpringerBriefs in Environmental Science, DOI 10.1007/978-3-319-59522-1_1

lack of knowledge and unsafe construction practices, earthquakes still cause widespread destruction leading to loss of lives and property. Earthquake can cause multi-level hazards resulting from ground motion, ground shaking, site effects, ground displacement, fire, floods, liquefaction, landslide and tsunami. When an earthquake hits, urban centres and cities experience more loss mostly because of high population density, improper planning, poor land use, substandard construction practices and quality of construction material.

What Is an Earthquake?

An earthquake is a sudden violent motion of the earth within a very limited region which lasts for a short time. The place where an earthquake originates inside the earth is called the focus (hypocenter) of the earthquake (Fig. 1.1). The point vertically above this, on the surface of the earth, is called its epicentre. Earthquake is associated with sudden rapid shaking of the earth causing fracture in the earth's crust, leading to release of strain energy stored in rocks which spread in all directions causing ground shaking (Fig. 1.2).

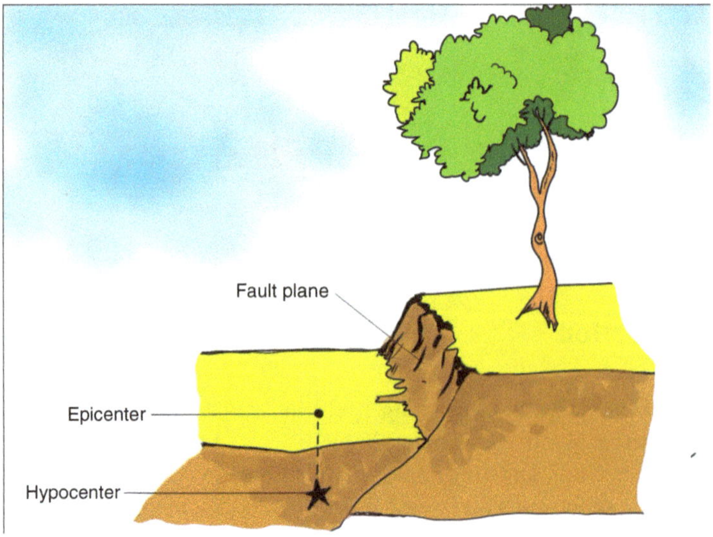

Fig. 1.1 The hypocenter, epicenter and the fault plane

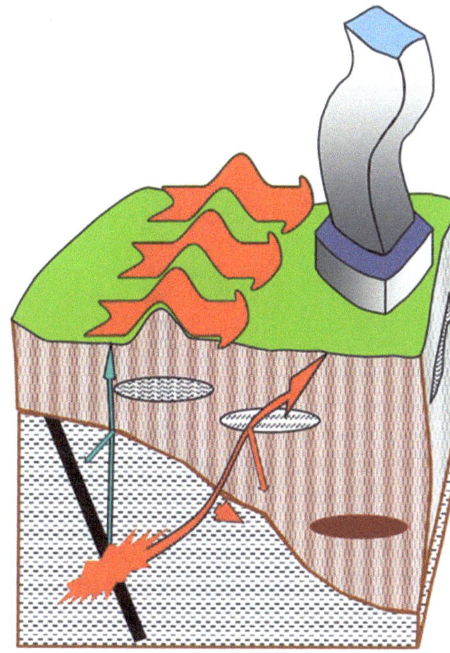

Fig. 1.2 Earthquake mechanism (*Source*: CVR. Murthy, IITK-BMTPC Earthquake Tip 2)

What Causes Earthquakes and Where Do They Happen?

The surface of the earth is made up of several wide, thin and rigid plate-like blocks. These are constantly rubbing against each other; and sometimes, their rubbing is strong enough to shake up the earth, release tremendous amounts of energy, and cause havoc. The current movement of the Indian plate is estimated to be around 50 mm/year. Of course, in our day to day lives, this kind of movement is negligible. However, a plate beneath the earth's surface, moving at this rate for a long enough time, can have a strong influence on the structure of the earth itself.

Earthquakes are more frequent in regions that lie directly above the edge of the plates. Volcanic activity and human activities like blasting, quarrying, mining and underground nuclear explosions can also cause quakes. Irrespective of the cause, enough precautions have to be taken to protect human dwellings when an earthquake strikes.

How Do Scientists Measure the Size of Earthquakes?

The size of an earthquake depends on the size of the fault and the amount of slip on the fault. Scientists use an instrument called the seismograph to measure the magnitude of an earthquake. Seismographs record the movement of vibrations beneath the earth, as zig-zag traces on a sheet of paper. Some seismographs are sensitive enough to detect earthquakes happening anywhere on earth. Figure 1.3a, b show seismograph and seismogram respectively.

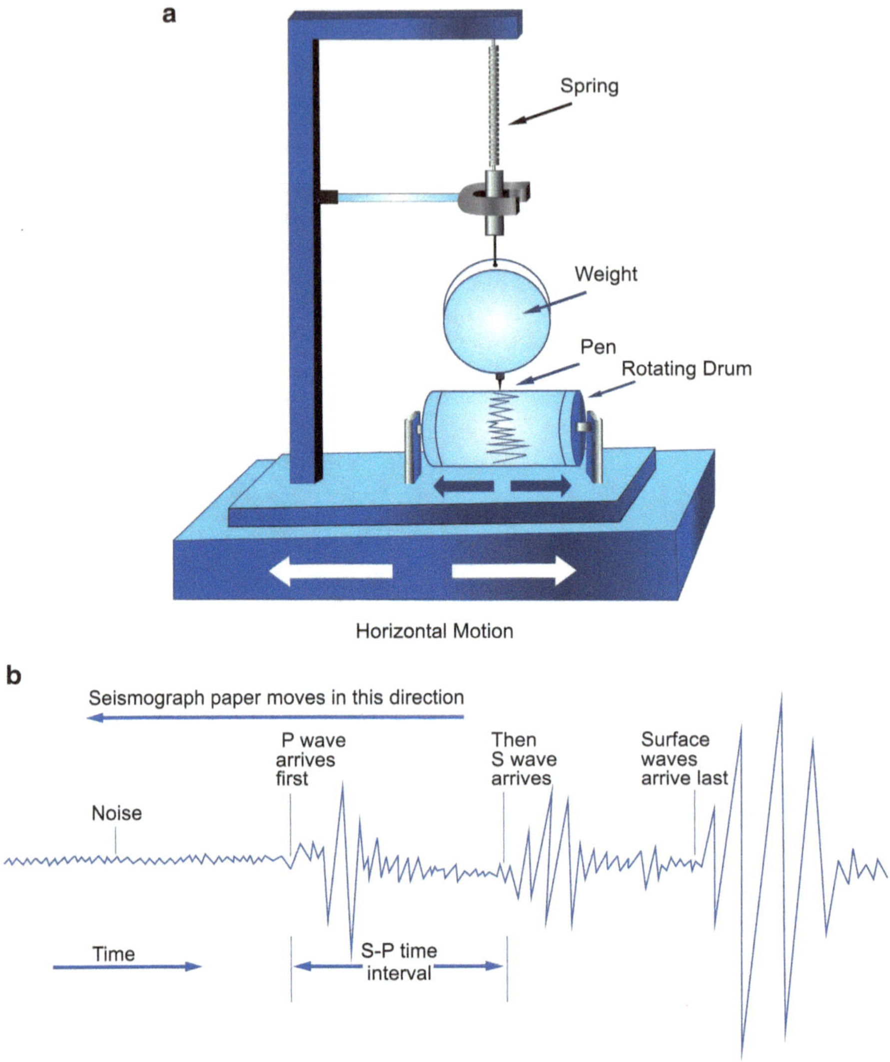

Fig. 1.3 (**a**) A seismograph and (**b**) a seismogram showing earthquake recording

A short wiggly line that doesn't wiggle very much means a small earthquake, and a long wiggly line that wiggles a lot means a large earthquake. The length of the wiggle depends on the size of the fault, and the size of the wiggle depends on the amount of slip. Earthquake size can be expressed qualitatively (non-instrumental) or quantitatively (instrumental). It is commonly expressed in terms of Intensity or Magnitude.

Intensity

Intensity is a measure of how strong the earthquake feels to an observer. It is the qualitative assessment of the kind of damage done by an earthquake. Intensity depends on distance to epicenter, strength of earthquake and local geology. It is determined from the intensity of shaking and the damage from the earthquake.

Magnitude

Magnitude is the quantitative measurement of the amount of energy released by an earthquake. Such a quantitative measure is needed to compare the size of earthquakes worldwide, which is independent of the density of population and the type of construction.

1.2 Earthquakes in and Around India

More than 50% of India's land area is vulnerable to earthquakes, and the Indian subcontinent has experienced earthquakes since ancient times. The northern region of India, which is along the boundary of the Indian plate and the Eurasian plate, is seismically very active. The north eastern movement of the Indian plate has caused deformation in the Himalayan region, Tibet and the North Eastern India. The North East Indian (NEI) region is known as one of the most seismically active regions in the world. However, the peninsular India, which is far away from the plate boundary, is a stable continental region, which is considered to be of moderate seismic activity. Even though the activity is considered to be moderate in the Peninsular India, one of the world's deadliest earthquakes occurred in this region (Bhuj earthquake 2001). The Table 1.1 below shows major earthquakes that triggered in and around India in last few decades (Fig. 1.4).

1.3 Seismotectonics of the Indian Subcontinent

The tectonic framework of the Indian subcontinent is spatio-temporally varied and complex. The rapid drifting of the Indian plate towards the Himalayas in the north eastern direction with a high velocity along with its low plate thickness (Kumar et al. 2007) might be the cause for the high seismicity of the Indian region. The Indian plate is moving northward at about 45 mm/year per year and it collides with the Eurasian Plate. Deformation within Asia reduces India's convergence with Tibet to approximately 18 mm/year, as Tibet is extending along East-West direction. This has resulted in the development of potential slip available to drive large thrust earthquakes beneath

Table 1.1 List of recorded high magnitude earthquakes in the Indian subcontinent (IMD)

No	Date	Lat	Long	Region	Magnitude
1	1819 Jun 16	24.0	70.0	Kutch, Gujarat	8.0
2	1869 Jan 10	24.5	92.5	Near Cachar, Assam	7.5
3	1885 May 30	34.1	74.8	Sopore, Jammu and Kashmir	7.0
4	1897 Jun 12	25.9	91.0	Shillong plateau	8.7
5	1905 Apr 4	32.3	76.3	Kangra, Himachal Pradesh	8.0
6	1918 Jul 8	24.5	91.0	Srimangal, Assam	7.6
7	1930 Jul 3	25.8	90.2	Dhubri, Assam	7.1
8	1934 Jan 15	26.6	86.8	Bihar–Nepal border	8.3
9	1941Jun 26	12.4	92.5	Andaman Islands	8.1
10	1943 Oct 23	26.8	94.0	Assam	7.2
11	1950 Aug 15	28.5	96.7	Arunachal Pradesh–China border	8.5
12	1956 Jul 21	23.3	70.2	Anjar, Gujarat	7.0
13	1967 Dec 11	17.4	73.7	Koyna, Maharashtra	6.5
14	1975 Jan 19	32.4	78.5	Kinnaur, Himachal Pradesh	6.2
15	1988 Aug 6	25.1	95.1	Manipur–Myanmar border	6.6
16	1988 Aug 21	26.7	86.6	Bihar-Nepal Border	6.4
17	1991 Oct 20	30.7	78.9	Uttarkashi, Uttrakhand	6.6
18	1993 Sep 30	18.1	76.6	Latur-Osmanabad, Mahrashtra	6.3
19	1997 May 22	23.1	80.1	Jabalpur, M.P.	6.0
20	1999 Mar 29	30.4	79.4	Chamoli, Uttarakhand	6.8
21	2001 Jan 26	23.4	70.3	Bhuj, Gujarat	7.7
22	2004 Dec 26	3.3	96.1	Off west coast of Sumatra	9.3
23	2005 Oct 8	34.5	73.1	Muzaffarabad	7.6
24	2011 Sep 18	27.8	88.1	Sikkim–Nepal Border	6.9
25	2015 Apr 25	28.18	84.7	Gorkha, Nepal	7.8

Fig. 1.4 Damage during 2001 Bhuj earthquake

the Himalaya. When continents converge, large amounts of shortening and thickening take place, like in the Himalayas and Tibet. Due to this massive collision, the Himalayas were formed and large numbers of earthquakes happen. A similar process, involving the Indian Plate and the Burmese micro-plate, results in earthquakes in the Andaman and Nicobar Islands. The plate boundary areas along the Himalayas and North East India are characterized by very high level of seismicity. In addition to this, there are earthquakes occurring within the Indian shield region, in the Indian peninsula and in adjoining parts of the Arabian Sea or the Bay of Bengal.

Geological Survey of India published Seismotectonic Atlas, SEISAT (2000) with details of linear seismic sources in India and adjoining areas with all available data related to earthquakes. It is a compilation of multi-thematic database comprising of 43 maps covering India and adjacent regions of neighboring countries on 1:1 million scale. Various details regarding geophysical, structural, seismicity and geothermal data relevant to seismotectonic activity are included in SEISAT. Figure 1.5 shows digitized map of linear seismic sources in and around India.

The seismic activity in India can be broadly characterized by three general seismotectonic considerations (Fig. 1.6): Tectonically active shallow crustal region, subduction zones and stable continental region. The subduction zone earthquakes can be further divided as regions with intraslab and interface earthquakes.

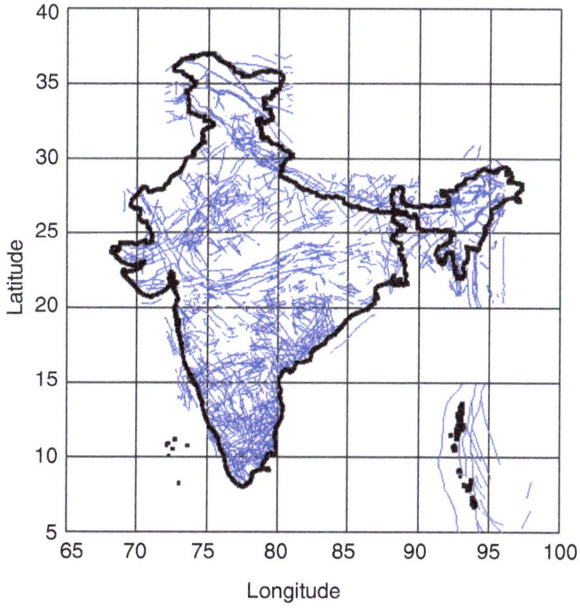

Fig. 1.5 Linear seismic sources identified in India (After SEISAT 2000)

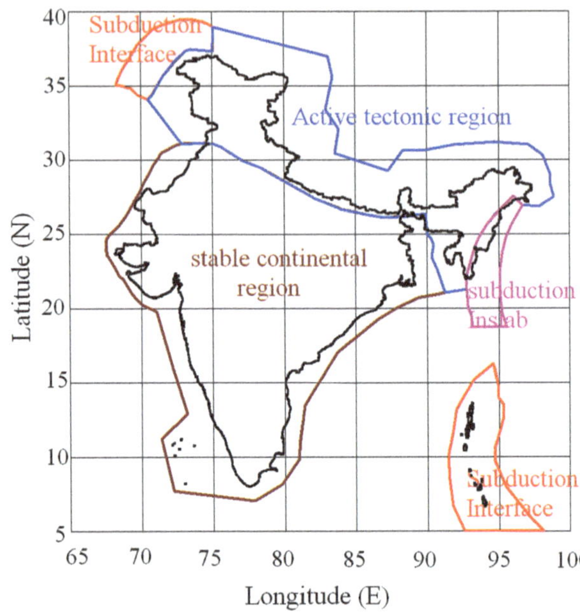

Fig. 1.6 Tectonic provinces in and around India (Kolathayar and Sitharam 2012a)

Active Tectonic Shallow Crustal Region

The seismicity of the Himalayan arc tectonic belt is connected with the under thrusting of the Indian plate beneath the Eurasian plate (Molnar and Tapponnier 1979; Krishnan 1953). The tectonically active interplate regions include the Himalayas and the southern Tibetan Plateau, northwest frontier province of Indian plate (Nath and Thingbaijam 2010; Kayal 2008). The movement of the Indian plate in the North Eastern direction and its collision with the Eurasian plate has created the most gigantic mountain range of the world—the Himalayas with an average height of 4600 m and the biggest and highest plateau region in the world—the Tibetan Plateau. The Indian plate was considered as one of the fastest moving plates in the world. Before its collision with the Eurasian plate, it had attained a very high velocity of around 20 cm/year (Kumar et al. 2007). The collision and the subsequent formation of the Himalayas and the Tibetan Plateau are associated with very high seismicity.

The entire North East Region is put under zone V of the Indian seismic zonation code (IS-1893, 2002). This region falls at the junction of N-S trending Burmese arc and E-W trending Himalayan Arc. Due to this, the entire region has suffered multiple phases of deformation processes and this has resulted in numerous geological structures.

Subduction Zones

The subduction zones include that of Hindukush- Pamir in the northwest frontier province, Indo-Myanmar arc, and Andaman-Sumatra seismic belt. The North East India, especially the region bordering China and Myanmar, is seismically very active. The seismicity of Indo-Burmese arc is related to the subduction of the Indian plate underneath the Southeast Asian plate due to north-eastward motion of India (Deshikachar 1974). The north-eastern corner of India, lying between the Himalayan and Burmese arcs, has a complex seismotectonic setup and very high seismic activity (Evans 1964). The earthquakes in this area are of intraslab in nature.

The Andaman Nicobar Islands (ANI), which is situated on the south eastern side of the Indian land mass, consists of about 527 islands. Andaman and Nicobar Islands (ANI) is the largest group of islands in the Bay of Bengal formed due to the thrust faulting between Indo-Australian plate and Burmese plate. These islands are branches of submarine mountains (Rai and Murty 2003). The southernmost tip of the Great Nicobar Island is only 150 km away from Sumatra, Indonesia. The region is regarded as one of the most seismically active regions in the world. The islands are associated with various geological issues such as subduction tectonics and earthquake processes, crustal deformation, active tectonics, volcanism, etc. The seismotectonic setting of the Andaman Nicobar Islands (ANI) was studied and well documented by various researchers (Eguchi et al. 1979; Mukhopadhyay 1984, 1988; Dasgupta and Mukhopadhyay 1993, 1997; Dasgupta et al. 2003; Curray 2005; Dasgupta et al. 2007a). The Andaman trench, west Andaman fault, Sumatra Fault System, Ranong Fault and Khlong Marui Fault are the active tectonic features in and around ANI (Dasgupta et al. 2003; Curray 2005). The entire island chain is along the plate boundary between the Indian plate and the Burmese plate. These regions come under subduction zones with interface earthquakes. The Andaman and Nicobar Islands, said to be located on a small tectonic plate which forms the ridges of the islands, is packed in between the Indo-Australian plate on the west side and the Eurasian plate in the north and the east (Dasgupta et al. 2007b). Spreading centres lie on the eastern side of the Islands and the Indian lithosphere on the western side subducts below the Andaman (Sunda) Plate making this region seismically active (Rajendran and Gupta 1989). Several damaging earthquakes and Tsunamis have hit the Andaman-Nicobar Islands in the past. The Sumatran earthquake of Dec 26, 2004 had also occurred along the same source and ANI was one of the worst affected regions during the Tsunami. ANI has been placed in zone V, the highest level of seismic hazard potential, according to the seismic zonation map of India (IS 1893-2002).

Stable Continental Shield Region

Peninsular India is delineated as Stable Continental Region (SCR) with low to moderate seismic activity (Chandra 1977). The seismicity of this region is of intraplate nature and appears to be associated with some local faults and weak zones (Rao and Murty 1970). The ENE–WSW trending Son-Narmada-Tapti zone is an important tectonic province in the northern edge of the peninsular shield. The major tectonic features in the southern part of the peninsula are the Deccan Volcanic Province, the Southern Indian Granulite Terrain, the Dharwar Craton, the Cuddapah Basin, the Godavari and the Mahanadi Grabens, and the Eastern and Western Ghats (Gupta

2006). The Bhuj earthquake (26 January 2001; Causality around 19,000) and the Latur earthquake (30 September 1993; Causality around 7928) are the deadliest earthquakes in this region. There were about ten earthquakes with magnitude 6.0 and above reported in this region.

1.4 A Glance Through Indian Seismic Design Codes

Seismic codes give guidelines to improve the performance of structures so that they may endure the earthquake effects without significant loss of life and property. Countries around the world have procedures outlined in seismic codes to help design engineers in the planning, designing, detailing and constructing of structures. Structures must be designed to withstand the forces and deformations induced during earthquakes.

Seismic codes are exclusive to a particular region, taking into account the regional seismology, accepted level of seismic risk, building types, and materials and methods used in construction. List of Indian standards dealing with Earthquake resistant construction are:

1. IS 1893 (Part 1): 2002 'Criteria for Earthquake Resistant Design of Structures: Part 1 General provisions and Buildings'
2. IS 1893 (Part 4): 2005 'Criteria for Earthquake Resistant Design of Structures: Part 4 Industrial Structures Including Stack Like Structures'
3. IS 4326:1993 'Earthquake Resistant Design and Construction of Buildings—Code of Practice'
4. IS 13827:1993 'Improving Earthquake Resistance of Earthen Buildings—Guidelines'
5. IS 13828:1993 'Improving Earthquake Resistance of Low Strength Masonry Buildings—Guidelines'
6. IS 13920:1993 'Ductile Detailing of Reinforced Concrete Structures Subjected to Seismic Forces—Code of Practice'
7. IS 13935:1993 'Repair and Seismic Strengthening of Buildings—Guidelines'

The regulations in these standards ensure that the structures are able to respond to earthquakes of moderate intensities without structural damage and of heavy intensities without total collapse. IS 1893 is the main code that provides the seismic zone map and specifies seismic design force. The seismic design force depends on the mass and seismic coefficient of the structure; the latter in turn depends on properties like the seismic zone in which the structure lies, the importance of the structure, its stiffness, the soil on which it rests, and its ductility.

IS 1893 illustrated a seismic zonation map in 1962 and revised it in 1966, 1970, 1984 and in 2002. The latest version of earthquake zoning map of India divides the country into four seismic zones (Zone II–V) as shown in Fig. 1.7, unlike its previous version which consisted of five or six zones in the country. According to the present zoning map, Zone V expects the highest level of seismicity whereas Zone II is associated with the lowest level of seismicity. There is immediate need to educate and

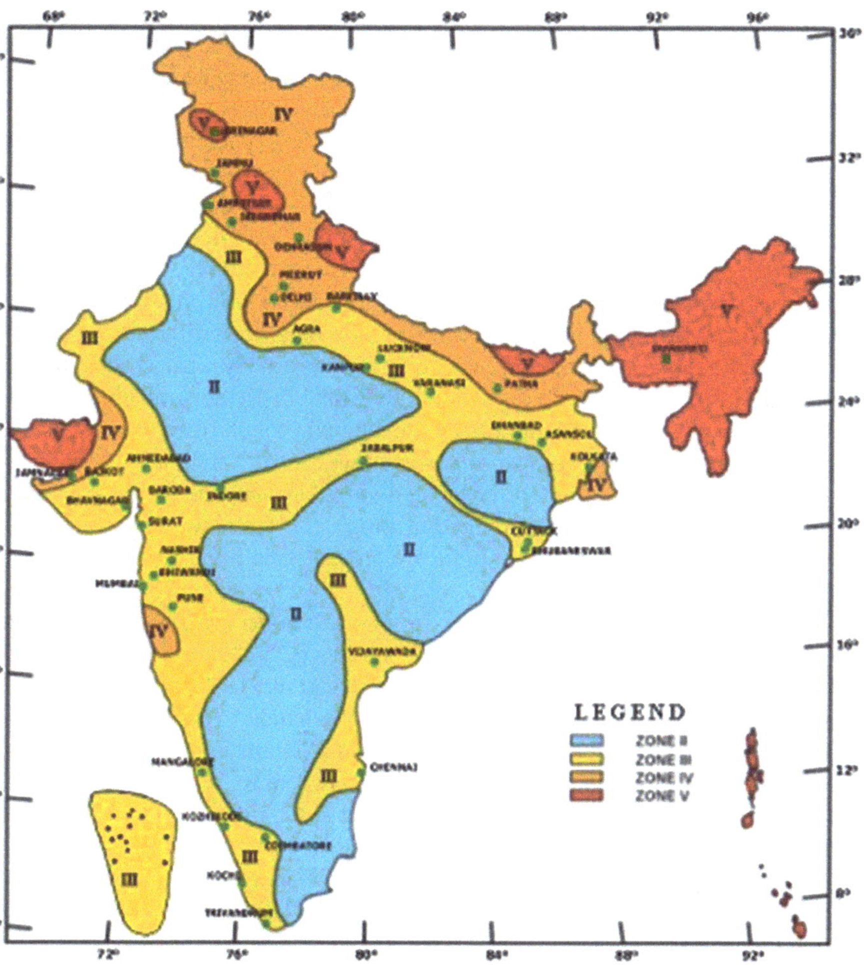

Fig. 1.7 Latest seismic zonation map of India (IS: 1893-2002)

empower the community residing in zones IV and V to face an eventual earthquake. The brief guidelines for constructing safe houses in the various areas lying in different seismic zones of the country and also about retrofitting etc. are given at the NIDM website http://nidm.gov.in/safety_earthquake.asp.

1.5 Seismic Hazard in India

The BIS-1893 (2002) delineates different seismic zones based on the past seismic activity and hence gets updated after a major earthquake event. For a developing country like India, the steps towards scientific seismic hazard evaluation are very essential to estimate an optimal and reliable value of possible earthquake ground motion during a specific time period. These predicted values will be an input to assess the seismic vulnerability of an area, based on which new construction and the restoration works of existing structures can be carried out.

Various researchers have attempted to evaluate the expected ground motion due to future earthquakes in and around India. Khattri et al. (1984) developed PGA hazard map with 10% annual probability of exceedance in 50 years and a similar was presented by Bhatia et al. (1999) as a part of Global Seismic Hazard Assessment Program (GSHAP). The deterministic seismic hazard map of entire India was prepared by Parvez et al. (2003). Kolathayar et al. (2012) estimated seismic hazard for India using deterministic approach with different source models and attenuation relations in a logic tree framework. Iyengar et al. (2010) and Nath and Thingbaijam (2012) developed the Probabilistic Seismic hazard map for Indian landmass. Sitharam et al. (2015) developed probabilistic maps for seismic hazard both at bedrock level and at surface level for the Indian land mass using different sets of attenuation relations to suit varied tectonic provinces in India and with different source models combined in a logic tree framework. Various seismic hazard maps developed for entire India are presented in Figs. 1.8, 1.9, 1.10, 1.11, 1.12, 1.13, and 1.14.

There were several other efforts by various researchers to estimate the seismic hazard for various isolated regions in the country using different methodologies. Few such studies specific to a region or city are Seeber et al. (1999) (Maharashtra state); Giardini et al. (1999) (Kolkata); Sharma (2003) (North India); Raghu Kant and Iyengar (2006) (Mumbai City); Sharma and Malik (2006) (North East India); Iyengar and Ghosh (2004) (Delhi); Jaiswal and Sinha (2007) (Peninsular India); Sitharam and Anbazhagan (2007) (Bangalore); Boominathan et al. (2008) (Chennai); Singh (2009) (Ahmedabad); Vipin et al. (2009) (South India); Mohanty and Walling (2008) (Kolkata); Menon et al. (2010) (Tamil Nadu); Kolathayar and Sitharam (2012b) (Anadaman Nicobar Islands), Abhishek et al. (2013) (Lucknow) and Sil et al. (2013) (North East India).

In addition, seismic microzonation work has been carried out in India in some of the important mega cities that have the potential of being damaged from future earthquakes. Microzonation is the process of subdivision of a region into number of zones based on earthquake effects in the local scale. Microzonation has generally been recognized as the most accepted tool in seismic hazard assessment and risk evaluation and it is defined as the zonation with respect to ground motion characteristics taking into account source and site conditions (TC4-ISSMGE 1999). Damage patterns of many recent earthquakes around the world, including the 1999 Chamoli and 2001 Bhuj earthquakes in India, have demonstrated that the soil conditions at a site can have a major effect on the level of ground shaking (Narayan et al. 2002). Seismic microzonation is the generic name for subdividing a region into individual areas having different potentials hazardous earthquake effects; defining their specific seismic behavior for engineering design and land-use planning. The role of geological and geotechnical data is becoming very important in the microzonation in particular the planning of city urban infrastructure, which can recognize, control

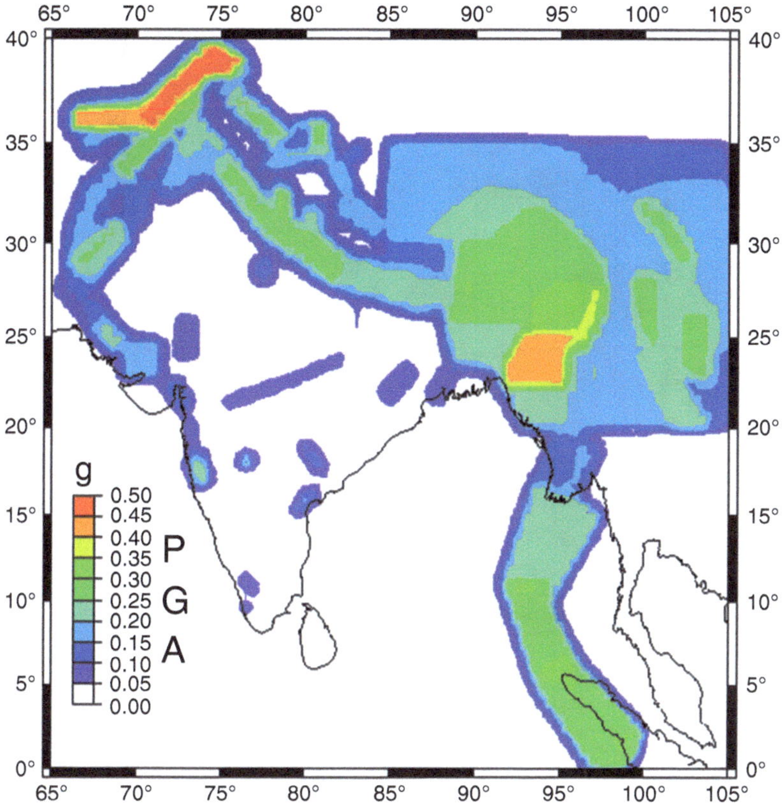

Fig. 1.8 The seismic hazard map of India for 10% probability of exceedance in 50 years (Bhatia et al. 1999)

and prevent geological hazards. The basis of microzonation is to model the rupture mechanism at the source of an earthquake, evaluate the propagation of waves through the earth to the top of bed rock, determine the effect of local soil profile and thus develop a hazard map indicating the vulnerability of the area to potential seismic hazard. Seismic microzonation will also help in designing buried lifelines such as tunnels, water and sewage lines, gas and oil lines, and power and communication lines. In order to understand the earthquake vulnerability of major urban centers, the Govt. of India has initiated microzonation of 63 cities in India after 2001 Bhuj earthquake. Many microzonation studies are under progress and some of them have been completed.

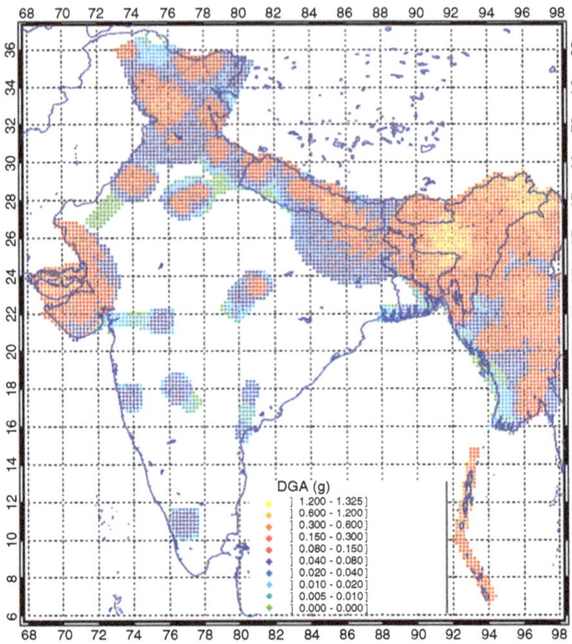

Fig. 1.9 The spatial variation of PHA (g) values given by Parvez et al. (2003)

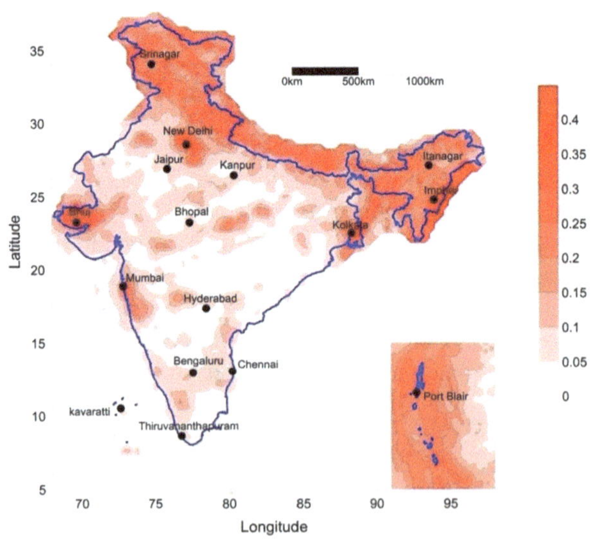

Fig. 1.10 Spatial variation of peak ground acceleration value (g) at bed rock level in and around India using DSHA (Kolathayar et al. 2012)

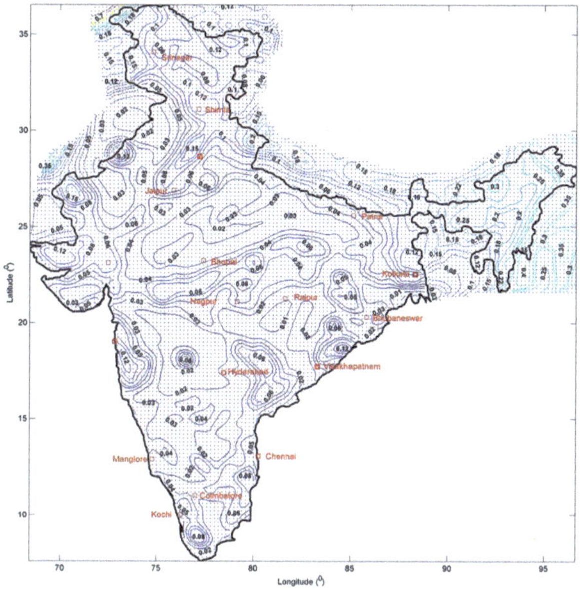

Fig. 1.11 PGA Contours with 10% probability of exceedance in 50 years (Return Period ~500 years) on A-type Sites (Iyengar et al. 2010)

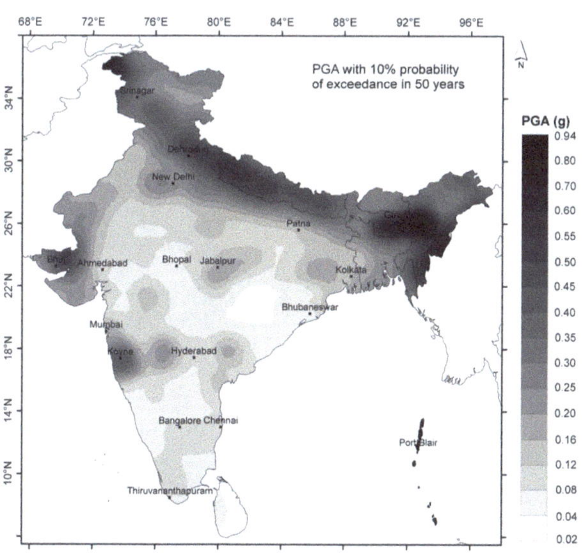

Fig. 1.12 Seismic hazard distribution in India in terms of PGA (Nath and Thingbaijam 2012)

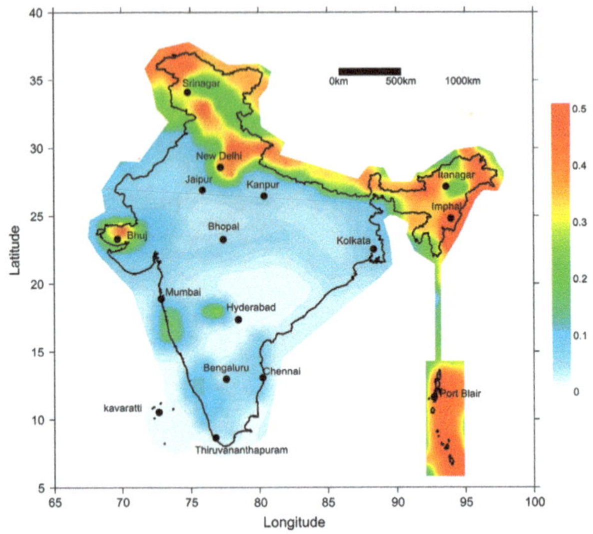

Fig. 1.13 PHA values (g) corresponding to a return period of 475 years (10% probability of exceedance in 50 years, PSHA) Sitharam and Kolathayar (2013)

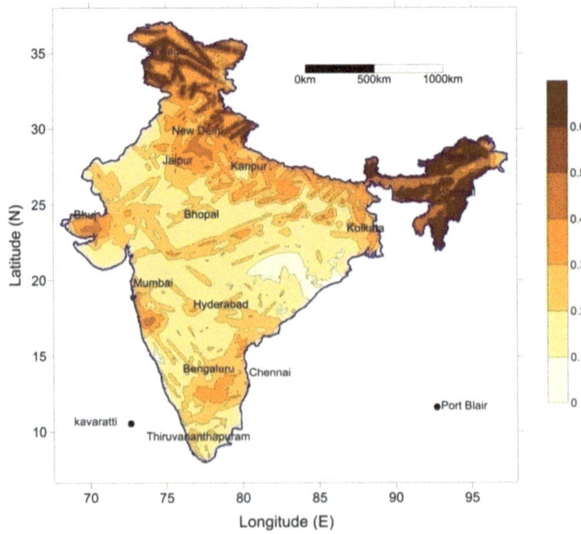

Fig. 1.14 Spatial variation in PHA (g) value at surface level throughout India (corresponding to 475 year return period, PSHA) Sitharam et al. (2015)

Chapter 2
Do's and Don'ts Before, During and After an Earthquake

Earthquakes never kill, Buildings do! Prepare before it is too late.

2.1 Introduction

Someday, an earthquake may strike the place you live in. Scientists cannot tell you when this will happen, whether it will be next year or during your grandchildren's lifetime. When that day comes, your community must be ready to face it. Seismic safety must not only be integrated into construction practices, but preparedness, emergency response, recovery, and long-term risk reduction efforts as well. It is a well-known fact that it is poorly planned infrastructures which cause disaster during an earthquake, not earthquakes as such. Identifying potential hazards in advance and appropriate planning can reduce the dangers of serious injury or loss of life from an earthquake.

T.G. Sitharam, S. Kolathayar, *Preparing for Earthquakes: Lessons for India*, SpringerBriefs in Environmental Science, DOI 10.1007/978-3-319-59522-1_2

2.2 Guidelines for Individuals and Families

Unlike other natural hazards, earthquakes strike suddenly without any warning. For people residing in regions vulnerable to earthquakes, there are certain measures that can be taken to reduce the chances of people being injured, and property being damaged. An earthquake does not cause death or injury by itself. People are hurt by collapsing building components or falling of heavy objects. Collapsing buildings and vibrations can cause short circuits and electric fires. All this leads to panic and confusion. With some precautions, it is possible to avoid such confusion.

Possible Measures Before an Earthquake

Undergoing a seismic evaluation is important to assess the earthquake resistance of one's home. Depending on when and how it was designed and built, the structure may have weaknesses that make it more susceptible to earthquakes. Common examples include structures not anchored to their foundations or having weak crawl space walls, unbraced pier-and-post foundations, or unreinforced masonry walls or foundations. Homes that are judged to have poor quake resistance should undergo seismic retrofitting with professional help. Before constructing, renting or buying a home, ensure that it complies with the seismic resistant provisions of building code for the particular region (Fig. 2.1).

Few points to remember:

- Fasten shelves securely to walls and place heavy objects on lower shelves.
- Store breakable items such as glass, and china pot etc. in low, closed cabinets.
- Hang items such as pictures or mirrors away from beds and places where people sit.
- Repair defective electrical wiring and leaky gas connections.
- Repair any deep cracks in ceilings or foundations.
- Get expert advice if there are signs of structural defects.
- Brace overhead light and fan fixtures.
- Identify safe places indoors and outdoors such as
  - Under strong dining table or bed
  - Against an inside wall
  - Away from where glass could shatter around windows, mirrors, pictures, or where heavy bookcases or other heavy furniture could fall over
  - In the open area away from buildings, trees, telephone and electrical lines, flyovers and bridges.
- Make sure all family members know how to respond after an earthquake.
- Teach all family members how and when to turn off gas, electricity, and water.
- Keep first aid kit ready in your home.
- Make a plan of where to meet your family after an earthquake.

Fig. 2.1 Possible scenario inside home during an earthquake

- In case family members get separated from one another during an earthquake, develop a plan for reuniting after the disaster. Ask a relative or friend in another district or state to serve as the family contact as it is easier to call long distance during a disaster.

During the Earthquake

If you are caught indoors at the time of an earthquake

- Be peaceful. Do not lose mental balance.
- Stay away from glass windows, doors, cupboards (almirahs), mirrors etc.
- Stay away from falling plaster, bricks or stones.
- Get under a table or a strong cot so that you are not hurt by falling objects.
- If open space is available nearby, go there.
- Do not use candles, matches or other open flames. Leaking gas or other flammable chemicals may accidentally light on fire.
- If trapped under debris, tap on a pipe or wall so rescuers can locate you. Use a whistle if available (Fig. 2.2).

If you are caught outdoors at the time of an earthquake:

- Keep away from tall chimneys, buildings, balconies and other projections.
- Check if you or anyone else is hurt. Use first aid at least on the cuts.
- Keep the streets clear for emergency services.
- A battery operated radio will help you to get important messages.
- Once in the open, stay there till the vibrations stops.

Fig. 2.2 If caught indoors in an earthquake, get under a strong table or cot

Fig. 2.3 If you are outside when an earthquake hits

Fig. 2.4 After an earthquake, check for damage in the building, save water, switch off gas supply and switch the mains off

- If you are in a moving vehicle, stop as quickly as possible (in a clear area) and stay in the vehicle. Once the shaking has stopped, proceed with caution. Avoid bridges or ramps that might have been damaged by the quake (Figs. 2.3 and 2.4).

What to Do After an Earthquake

- Check yourself for injuries.
- Help injured or trapped persons. Don't try to move seriously injured people unless they are in immediate danger of further injury. Wait for medical help to arrive.
- While entering buildings, be extremely cautious. Building damage may have occurred where you least expect it. Carefully watch every step you take.
- Examine walls, floor, doors, staircases and windows to make sure that the building is not in danger of collapsing.
- Look for and extinguish small fires. Eliminate fire hazards. Putting out small fires quickly, using available resources, will prevent them from spreading. Fire is the most common hazard following earthquakes.

- Check for gas leaks. If you smell gas or hear a blowing or hissing noise, open a window and quickly leave the building. Turn off the gas, using the outside main valve if you can, and call the gas company from a neighbours' home.
- Clean up spilled medicines, bleaches, gasoline, or other flammable liquids immediately.
- Open the closet and cabinet doors cautiously. Contents may have shifted because of shaking during the earthquake and could fall, creating further damage or injury.
- Inspect your home for damage. Get everyone out if your home is unsafe. Aftershocks following earthquakes can cause further damage to unstable buildings. If your home has experienced damage, get out before aftershocks happen.
- Help neighbours who may require special assistance. Elderly people and people with disabilities may require additional assistance. People who care for them or who have large families may need additional assistance in emergency situations.
- Listen to a portable, battery-operated radio (or television) for updated emergency information and instructions. If the electricity is out, this may be your main source of information. Local radio and local officials provide the most appropriate advice for your particular situation.
- Watch out for fallen power lines or broken gas lines, and stay out of damaged areas. Hazards caused by earthquakes are often difficult to see, and you could be easily injured.
- Stay out of damaged buildings. If you are away from home, return only when authorities say it is safe. Damaged buildings may be destroyed by aftershocks following the main quake.
- Check for sewage and water line damage. If you suspect sewage lines are damaged, avoid using the toilets and call a plumber.
- Watch animals closely. Tie dogs and place them in a fenced yard. The behaviour of pets may change dramatically after an earthquake. Normally quiet and friendly cats and dogs may become aggressive or defensive.
- Do not spread and believe rumours.
- Do not waste water. Do not crowd around damaged areas or buildings.
- Stay as safe as possible during an earthquake. Although smaller than the main shock, aftershocks cause additional damage and may bring weakened structure down. Aftershocks can occur in the first hours, days, weeks, or even months after the quake.

2.3 School Emergency Plan

At schools, regular earthquake drills should be performed to prepare students for an evacuation in the event of an earthquake.

Fig. 2.5 Mock drill at school on what to do during an earthquake

1. The school should have a system set up to communicate with families during a crisis.
2. The school must store adequate food, water, and other basic supplies.
3. The school should be prepared to be a shelter-in-place if need be, and a safe place near should be identified, to go if they must get away.
4. In case where schools institute procedures to shelter-in-place, parents may not be permitted to drive to the school to pick up their children. They should monitor local media outlets for announcements about changes in school openings and closings, and follow the directions of local emergency officials (Fig. 2.5).

Model school designs for construction in various seismic zones in India were published by the Govt. of India under UNDP Disaster Risk Management Program in 2006 and Guidelines for School during various disasters were published by NIDM, which can be accessed from below links.

http://nidm.gov.in/PDF/safety/school/link2.pdf

are available at http://nidm.gov.in/safety_school.asp

2.4 Approach for Business Organizations

Business organizations located in earthquake prone areas can do many things to reduce the loss of life and property, and to ensure least disturbance to day-to-day function if an earthquake occurs. Workplace preparedness requires the participation of owners, managers, workers, and designers of the buildings.

Before the Earthquake

Following are activities that can be undertaken before the earthquake:

1. Check with local building-regulatory agency to find out whether, and for how long, structures in the area have been subject to building codes containing seismic design provisions.
2. Make sure that all non-structural items are effectively anchored, braced, reinforced, or otherwise secured. Design and construction professionals should properly secure some of these components, while others can be made safe by maintenance staff or other employees.
3. Every employee, from top managers to part-time and temporary workers, needs to learn what to do during an earthquake. Safety orientations should emphasize safe places to drop, cover, and hold on during earthquake shaking and safe locations where people can meet after the shaking has stopped.
4. Conduct regular compulsory earthquake drills to employees for them to practice what they know about what to do during earthquake shaking and to react spontaneously and safely. To help protect workers in the immediate aftermath of earthquakes or other disasters, arrange for employees to be trained in first aid and the use of fire extinguishers. Earthquakes should be thoroughly integrated into the organization's emergency preparedness, response, and recovery planning.

The employers can play an important role to contribute to the well-being of the communities from which they recruit employees, clients, and customers. There are many ways that businesses organizations can help strengthen the disaster resilience of their communities. Some of these ways include promoting preparedness among suppliers, clients, and other business contacts; and sponsoring or participating in local earthquake drills, preparedness events, or awareness and education campaigns.

During the Earthquake

When earthquake shaking begins, the employees should respond calmly applying what they have learned about 'Do's and Don'ts during an Earthquake'.

After the Earthquake

Once the shaking stops, workplaces should be prepared to implement pre-arranged, earthquake-specific emergency response and recovery plans. Employees must keep in mind that aftershocks may strike at any time; exacerbating hazards created by earlier shaking and requiring that everyone again drop, cover, and hold on.

Regardless of the severity of this earthquake, learn from the experience. If structures must be repaired or rebuilt, use this opportunity to ensure that weaknesses are addressed and that new facilities are compliant with seismic building standards. If building contents were damaged, improve how such items are secured. If employees were insufficiently prepared to react safely, safety training and the frequency of drills must be increased.

2.5 Note for Public Policy Makers

Current Status

The Central Government has enacted Disaster Management Act 2005 which provides institutional mechanism for responding to disasters including earthquake. The National Disaster Management Authority (NDMA) has been constituted which is broadly mandated with preparation of guidelines for management of disasters. NDMA has already published the guidelines on management of earthquakes and conducted few mock drills for earthquakes. The Central Government has also raised National Disaster Response Force (NDRF) and equipped NDRF with Collapse Structures Search and Rescue (CSSR) equipment for responding to earthquakes.

NDMA Guidelines (2007) listed critical areas of concern for the management of earthquakes in India and addressed various techno-legal and techno-financial issues pertaining to earthquake disaster management. Therein timeframes have been given for Important Milestones for the Implementation of the Guidelines. The activities were envisaged to be implemented in two phases. Phase to be concluded by 31 December 2008 and be reviewed by 30 June 2009. Based on this review, the activities in Phase II were supposed to be designed in the second half of 2009 and Phase II to begin by 1 January 2010. It is imperative that these milestones are implemented at the earliest.

National Centre for Seismology maintains operation of a countrywide national seismological network, to detect and locate earthquakes occurring in and around the country as well as dissemination of earthquake information to the user agencies including media in least possible time to help take appropriate follow up actions. The seismic microzonation studies of some of the major cities in the country such as Jabalpur, Guwahati, Bangalore, Sikkim, Ahmedabad, Gandhidham-Kandla, Kolkata and Delhi has been completed. As a part of 12th 5 year plan, a project has been initiated for seismic microzonation of 30 more cities. The seismic microzonation maps/results are used by various agencies as mitigative measures for design and construction of earthquake resistant structures, land use planning and for enacting building bye-laws towards minimizing damage to property and loss of lives due to earthquakes.

In addition, efforts are being made to develop an Early Earthquake Warning system. In this direction, a pilot study has been initiated through Department of

Earthquake Engineering, IIT Roorkee. In this project, one hundred sensors will be installed in Uttarakhand, which will be connected with central receiving station. One such system is tested and has become operational which can provide a lead time of about 30–60 s for issuing warnings to cities like Delhi, located about 300 km from Himalayan source. Such warning, if successful, can enable people to move to safer places and emergency shutdown of critical installations, to an extent.

Earthquake resistant building construction codes have already been published by the Bureau of Indian Standards, and are updated from time to time.

Way Forward

Seismic activity in India is spatially varied and complex with various tectonic provinces of different characteristics. Also, India is a very large country with a huge population. Hence there is a need for decentralization for effective management of earthquakes, empowering local bodies to take needful actions in accordance with the seismic vulnerability of respective regions.

1. Public awareness should be enhanced through mass communication, media houses and by all means. It is recommended that the Prime Minister himself take lead in the awareness campaign for Earthquake Preparedness like it was done for 'Swachh Bharat' campaign.
2. For apartment buildings and other such dwellings, earthquake preparations should be made through the owners' association, including appropriate seismic evaluations, earthquake-proofing, repairs, and functional upgrades through renovation.
3. Important buildings such as schools and hospitals which are needed to secure emergency shelter and medical functions should receive financial and technical support to proceed with earthquake proofing efforts.
4. Measures should be advanced to prevent the fall of non-structural elements such as ceiling materials and lighting fixtures in schools, etc.
5. Extend financial and technical support needed for earthquake proofing and rebuilding of apartment buildings.
6. Promote public awareness of regional risk levels, information on potential damage and examples of earthquake proofing initiatives.
7. Work to encourage the spread of multiunit housing equipped to enable continued living even in the event of a disaster.
8. By systematically conducting seismic retrofitting and rebuilding, advance the earthquake proofing of publicly-owned housing.
9. Encourage the rebuilding and upgrade of ageing buildings through urban re-development.
10. Indian agencies can investigate how well the advanced warning system worked for the Great East Japan earthquake of 2011 and consider investing in a similar

system for areas of Kashmir, Himalayan region, Andaman Nicobar, and the North East India.
11. We need to look at how Japan has prepared for the Great Tokai earthquake as a model for what to do in its most earthquake-prone areas. India needs to expand this model to seismically active regions of India.
12. Encourage State Governments to actively participate in Earthquake disaster preparedness plan. State Governments should involve local bodies like corporations, Municipalities and Panchayats to carry out the plans effectively and systematically reaching out to all regions vulnerable to earthquakes.
13. Earthquake safety aspects should be incorporated in the building by-laws and it should be made mandatory. (e.g.: Gujarat State Disaster Management Authority).
14. Usually Municipality or corporation approve the design of buildings. The problem faced by municipalities or corporations is that they do not have competent staff. They can take help from competent people from academic institutions. There should be provision for fund for municipalities to pay for them.
15. There is a need to create awareness and confidence among architects and town planners. Open space is required for rehabilitation which should be looked upon by town planners.
16. Normally, resistance towards incorporating earthquake safety features come from architects, builders and clients because of extra cost and effort. There is a need to educate them that life is more important than extra cost.
17. With regard to Design and Details, in most of the cases, the earthquake safety features are not given in distributed drawings. It should be made very clear so that a site engineer should be able to follow. Supervisors, site engineers and craftsmen should be trained. This is especially relevant in rural areas.
18. Construction activities in rural areas must be given special attention.
19. All parts of IS 1893 should be made available at the earliest.
20. Educational institutions starting from Kindergarten should be made aware of earthquake safety and basic safety aspects should get into school books. Mock drills must be conducted in schools situated in vulnerable regions.

The country needs a comprehensive law and policy that address all aspects of disaster management under one roof to balance mitigation and preparedness efforts more evenly.

2.6 Restoration Strategy

The following guiding principles can form the basis of strategy and planning of post disaster recovery:

1. Participation of the community by empowering them to take control of reconstruction of their own houses and ensuring facilitation of owner-driven reconstruction.

2. Housing reconstruction must include holistic habitat development including basic services and community set-up.
3. Reconstruction should be a platform to build long-term community resilience, strengthening community capacities to mitigate future disasters by improving construction practices for majority of the buildings in the country.
4. Strengthening the local economy through practices favorable to the poor, marginalized and informal sector. Reconstruction should provide an opportunity for the poor to upgrade their living conditions.
5. Ensuring sustainable reconstruction processes, taking account of climate change, natural resource management and scientific risk assessments.

Chapter 3
Resilient Community for Earthquake Preparedness

Prepare because you care. Safe sail equals safe life!

3.1 Introduction

Earthquake preparedness is not a part of the public consciousness even in the vulnerable regions of India. There are many things that individuals and communities residing in seismically active regions can do that will reduce the havoc and loss during a quake. These steps can be a complementary measure, alongside other activities that focus on the overall development of the community; with a team

T.G. Sitharam, S. Kolathayar, *Preparing for Earthquakes: Lessons for India*,
SpringerBriefs in Environmental Science, DOI 10.1007/978-3-319-59522-1_3

spirit. Usually the aftermaths of a disaster bring solidarity among the members of the community. There is a need to bring this spirit of togetherness and solidarity well before a disaster so that a community can be prepared well to face the disaster and thus reduce the loss of life and properties.

Common arguments against preparing for disasters include:

- It will harm the economy with adverse effects on business and tourism
- People will panic if future hazard is discussed openly
- The community is too poor to prepare for earthquakes, and
- There are too many other high priority problems in the community

Worldwide experience shows that none of these arguments are valid. Some of the world's strongest economies and favorite tourist destinations, have been publicly discussing and planning earthquake preparedness for decades with no harm to business or tourism. Hundreds of communities can point to experiences in which openly focusing on how to reduce disaster risk built confidence in their population and business community, rather than causing problems. Poor communities around the world have focused on disaster preparedness: many risk-reducing activities require people-power, not financial resources. People often think of emergency planning and preparedness as the government's job, but the best prepared communities involve every sector of society in disaster planning. Earthquake safety programs need to affect everyone in a community, which requires involving a wide range of people from every element of society in developing those programs.

Anyone can be an activist for earthquake safety and preparedness: be it government officials, business leaders, members of community organizations, or any concerned citizens. All that is required is a commitment to making your community safer and a willingness to learn. People from every sector of society have critical roles to play in earthquake safety. Some aspects of earthquake preparedness generally need to be led by governments, such as developing official warning systems and evacuation plans as discussed in earlier sections. Other equally critical preparedness activities can be led by community members outside of the government. Someday, an earthquake may strike your community. Scientists cannot tell you when this will happen, whether it will be next year or in your grandchildren's lifetime, for example. When that day comes, you want your community to be ready.

3.2 Form a Good Team

The people whom you should consider involving in your preparedness include:

1. Emergency responders: police, fire fighters, medical personnel, safety officials, departments with emergency management roles, agencies with expertise on geological hazard
2. Political leadership: mayors, panchayat presidents, MLAs etc.
3. Local media

4. Local non-governmental organizations with emergency response directives or expertise.
5. Women's groups
6. Representatives of disadvantaged groups likely to be strongly affected by earthquakes (poor, disabled, elderly)
7. Respected local figures: religious leaders, professors, representatives of professional associations (engineering, architecture, medicine), etc.
8. Businesses leaders and representatives from tourism and hotel industries
9. Management of nearby schools and colleges
10. Non-Governmental Organizations those are active in the community

After forming a good team, develop a structure that is suitable for your community and the resources available. The level of structure of teams can vary significantly, from lose networks of volunteers with advisory committees to legally integrated organizations. Different organizational forms make sense at different times during the preparedness efforts.

3.3 Understand the Hazard in the Locality

It is important that the team is aware of the level of seismic hazard potential at their locality. Best way to learn is to refer scientific seismic hazard maps prepared by experts (Refer Chap. 1 of this document). These hazard maps can give a broad idea on the expected risk at your region. Conduct workshops, seminars and regular meetings discussing about the hazard at your region so that the idea of earthquake is transmitted to public consciousness. This should be done without inculcating a sense of with fear but with a positive outlook that working together towards earthquake preparedness can make the community ready to face any disaster. This will bring a spirit of togetherness and vibrancy among the people which may pave way for other developmental activities in the community as well.

3.4 Create Awareness

It is important to remember that disasters like earthquakes destroy all progress a community has been making in every area. With time and persistence of campaigners, community leaders will learn that earthquake preparedness is not just good for the community, it is essential to its long-term survival and stability. This awareness will slowly transmit through generations.

The community can carry out lot of preparatory planning after discussion with the members of the community.

1. A special section in the local newspaper can be published with emergency information on earthquakes. The information can be localized by printing the phone numbers of local emergency services offices and hospitals.
2. A week-long series on locating hazards in the home can be conducted.
3. The information on what to do during an earthquake can be distributed to individuals and families.
4. Tips on conducting earthquake drills in the home can be provided.
5. Community can identify or even build open playground which can serve dual purposes. The ground can be used for games or social functions during safe period and the same ground can act as a shelter for people to camp in the event of a disaster or to land helicopters during emergency.
6. Interview representatives of the gas, electric, and water companies about shutting off utilities.

Work together in your community to apply your knowledge to building codes, retrofitting program, hazard hunts, and family emergency plans.

When respected, trustworthy and influential people in your community take earthquake preparedness seriously, others are inspired to be prepared. It's a good idea to get this type of 'inspirational' person involved in the outreach efforts (Figs. 3.1 and 3.2).

Organize Community Outreach Events

Keep events small and focused on local concerns.

Encourage community members to participate through providing a range of activities, questions and discussions. People learn better through activities and tend not to learn much from lectures. Activities should be led, whenever possible, by people who are from or who are well-known in the local community where you are focusing.

- Community meetings
- Workshops
- Emergency planning for schools
- Emergency planning for hospitals
- Community earthquake drills
- School evacuation drills
- Formation of Student safety clubs
- Neighborhood preparedness teams
- Rallies or marches for preparedness
- Displays or programs at community fairs and festivals
- Presentations to community clubs and organizations
- Street plays
- Discussions/Talks by Earthquake survivors
- Earthquake signs

Fig. 3.1 Village level camps to create awareness on earthquake preparedness

Fig. 3.2 Community awareness using electrodynamic shaker with a slip table

- Memorials to past earthquakes
- Door-to-door education campaigns
- Newspaper articles
- Radio or television documentaries
- School curricula covering earthquake preparedness
- Web pages
- Opening of Disaster risk information centre

3.5 Review the Plan and Improve the Efforts

Reviewing the success of the programs in preparing community members will allow to learn from mistakes and to escalate the effectiveness of future programs. Some ideas on how to evaluate the programs include:

- Distribute questionnaires at events such as community meetings or drills. How did people learn about the event? Did people learn from it? What aspects did they find most useful?
- Collect informal feedback from people working with the team. Everyone who attends community meetings or interacts with the public in other ways will have thoughts about how people are responding to the preparedness efforts; what they understand and misinterpret, and whether people will be ready to prepare for the earthquake.
- Conduct formal polling on awareness and preparedness rates in the community before and after the outreach efforts. This approach, while most expensive, gives the most objective feedback on how successful the outreach program has been.

3.6 Keep Preparedness Alive for Long-Term

The earthquake preparedness aspects must peep into people's mind and must continue throughout generations to create a well prepared and alert community. Persuading people in this regard takes time. People, governments and other groups need to be continually reminded to keep their preparedness efforts up-to-date and ongoing. Every family should update their knowledge and preparedness plans based on changes in where they work, live or go to school.

Governments need to update their plans to accommodate changes in the community's growth. The new members entering the community should be educated.

Some approaches that communities can keep disaster preparedness activities ongoing include:

- Create an organization to focus on disaster preparedness issues. The structure of an organization spreads the momentum for preparedness activities away from individuals or temporary groups to a permanent entity.

- Integrate earthquake preparedness into government programs. Government officials can make sure that development activities are earthquake-resistant.
- Integrate earthquake preparedness into programs of other institutions, such as schools and businesses. Schools can add earthquake preparedness to their curricula.

Communities can survive earthquakes if they prepare. How important the public and policy makers feel about the need for earthquake preparedness at any given moment will fluctuate over time, based on world events and local issues. As an activist, one should be ready to take advantage of events that raise the publics' interest in earthquakes. These events could be large earthquakes elsewhere in the world, or other types of natural disasters. These events can be as opportunities to conduct community awareness events. When an earthquake hits, we should be ready not only to survive but to bounce back to be more lively with great hopes. This makes our efforts meaningful.

Chapter 4
Earthquake Readiness Index Tool for India

Know Your Readiness Index and Boost Preparedness Level

This chapter presents the development of a survey based tool specific to India, to assess the earthquake preparedness level of individual, families and communities residing in earthquake prone regions.

4.1 Questionnaire Development

The development of questionnaire for assessing the Earthquake Readiness Indices (ERI) at individual and community levels involved the following steps:

1. Literature survey
2. Focus group discussions
3. Sample questionnaire
4. Establishing face validity and content validity of sample questionnaire
5. Pilot survey
6. Measuring construct validity of sample questionnaire
7. Measuring reliability of sample questionnaire

Literature Survey

A literature review was carried out as the first step towards the questionnaire development. A detailed analysis of existing literature on assessing earthquake preparation levels of public, on a global scale is presented in Table 4.1.Various online sources were also examined for obtaining precise and accurate information on

T.G. Sitharam, S. Kolathayar, *Preparing for Earthquakes: Lessons for India*,
SpringerBriefs in Environmental Science, DOI 10.1007/978-3-319-59522-1_4

Table 4.1 List of studies on earthquake preparation measures

Study	Sample	Earthquake preparation measure
Ardalan and Sohrabizadeh (2016)	450 Households of Iran	18 item Earthquake Readiness Scale
Baytiyeh (2014)	378 college students of various disciplines from Lebanon	Unspecified no: of questions
Becker et al. 2012	Residents of 3 communities Napier, New Zealand, Timaru and Wanganui.	Unspecified no: of questions
Yang et al. (2010)	110 Graduate Students in Wuhan University	Unspecified no: of questions
Becker (2010)	1500 Residents of Timaru	34 item Questionnaire
Paton et al. (2010)	255 Residents of New Zealand, Napier and 251 from Japan and Kyoto.	34 item Questionnaire
Okazaki et al. (2008)	800 Households Each in Indonesia, Nepal , Pakistan and Turkey	35 item Questionnaire
Spittal et al. (2008)	358 residents of Wellington, New Zealand	43-item measure
McClure et al. (2009)	240 Citizens of Willington, New Zealand	Unspecified no: of questions
McClure et al. (2007)	400 Companies of Wellington	Unspecified no: of questions
McClure and Allen (2001)	100 students and 124 residents of Wellington for Study 1 and Study 2 respectively	5 item earthquake preparation checklist
Hurnen and McClure (1997)	96 Wellington citizens	12 item checklist
Russell and Bourque (1995)	690 residents and 656 residents interviewed after Whittier Narrows earthquake and the Loma Prieta earthquake respectively	17 item scale
Edwards (1993)	544 Tennessee residents	14 item checklist
Farley et al. (1993)	559 households from three cities in Montana	4 item checklist
Dooley et al. (1992)	1600 California residents	5 item checklist
Mileti and O'Brien (1992)	1652 residents of Santa Cruz and San Francisco	6 item earthquake preparedness checklist
Mulilis and Lippa (1990)	114 homeowners in California	27 item Mullis–Lippa Earthquake Preparedness Scale
Mulilis et al. 1990	154 homeowners, and 296 students of California	27 item Mullis–Lippa Earthquake Preparedness Scale
De Man and Simpson-Housley (1987)	130 residents of San Francisco	Unspecified no: of questions
Hirose (1986)	1018 Japanese participants	13 item checklist

(continued)

Table 4.1 (continued)

Study	Sample	Earthquake preparation measure
Turner et al. (1986)	1450 participants from California	16 measures for earthquake preparedness
Hirose and Ishizuka (1983)	1018 Japanese participants	13 item checklist
Kiecolt and Nigg (1982)	1450 residents of California	Intention to change current location of residence in the next 5 years
Jackson (1981)	302 householders from British Columbia and California.	Open ended questionnaire
Sullivan et al. (1977)	1400 residents from California	Insurance purchase
Jackson and Mukerjee (1974)	120 residents from San Francisco	8 item checklist

earthquake preparation and questionnaire development. Information regarding the different steps and measures to prepare for an earthquake and lists of recommendations made by several disaster management agencies and guidelines on questionnaire development were studied.

Focus Group Discussions

A few focus group discussions were carried out with participants from various disciplines who had direct and indirect experience with disaster, specialists in earthquake studies, medicine and marketing. This gave a better idea about the factors that are to be considered with regard to the framing of questionnaire accordingly with the precise content in line with the aim of the questionnaire development.

Sample Questionnaire

A sample questionnaire with the basic framework and primary contents derived from the above mentioned scales, information from websites and focus group discussions was developed. It included a 30-item scale for ERI for individuals, 20-item scale for ERI for community along with demographics for interpretative studies.

The 30-item scale included 30 steps/measures that were considered necessary for an individual to safeguard life and property against a major earthquake. This scale was developed keeping the key factor as an individual's level of intention to prepare against a major earthquake. Three point Likert scale was employed in this 30-item scale. The 22 item scale was split into three keeping the key factors as community participation, collective efficacy along with trust, and empowerment. It employed three point Likert scale, five point Likert scale and three point Likert scale respectively.

Establishing Face Validity and Content Validity of Sample Questionnaire

Subject experts established the face validity and content validity of the questionnaire. The process was carried out through self-completion questionnaire which were emailed. Subsequent revision was made in the sample questionnaire making it 27 item scale for ERI for individuals and 20 item scale for ERI for community. The revised questionnaire was then sent for pilot survey.

Pilot Survey

The sample questionnaire was pretested by sending it for pilot survey using Google Forms. The sample size obtained for pilot survey was 132 which fulfilled the bare minimum criteria of 3 respondents against 1 item in both the scales. The completion rate of the survey was 99.38%. The questionnaire was reconstructed considering the feedback from the respondents.

Measuring Construct Validity of Sample Questionnaire

Construct validity was determined by exploratory factor analysis. It was done by clustering items into common factors. Principal component analysis (PCA) and Varimax methods were employed for FEA. To ensure an appropriate sample size for running factor analysis, the Kaiser-Meyer-Olikin (KMO) sampling adequacy was computed. The accepting values of ≥0.5 were recommended. The KMO sampling adequacy of the tool was 0.7 indicating the sample size of 132 had been reasonable for performing factor analysis for the pilot study.

Measuring Reliability of Sample Questionnaire

Internal consistency reliability check produced alpha co-efficient (Cronbach's alpha) greater than 0.7, indicating acceptable correlation between the items of the questionnaire. Test-retest reliability was established by computing Spearman's correlation, which ranged between 0.8–1 indicating good to excellent reliability. The retest was performed among 50 respondents of pilot survey after two months interval. Based on the above-mentioned steps the finalised questionnaire for field survey was developed (Appendix 1).

4.2 Significance of ERI

The score obtained for ERI for individuals gives the readiness of an individual against a major earthquake. The statements in the 27-item scale are the measures to be adopted by an individual for being earthquake ready. The score for obtained ERI for community gives the readiness of a community against a major earthquake. This scale can be used anywhere in India by anyone at any time to assess their readiness level against a major earthquake. The 20-item scale can be used by any organisations or agencies to assess the level of readiness of any community at any time against a major earthquake.

4.3 Significance of Demographics

Demographics included in the questionnaire helps in quantifying the obtained score in terms of various demographic parameters such as gender, age, educational qualification, employment status, annual income etc. It also gives a better picture about the number of vulnerable members in a household or community.

4.4 Confirmation of Validity and Reliability of Questionnaires

An extensive survey with detailed analysis was carried out at the city of Coimbatore for the confirmation of validity and reliability of the questionnaire. Cronbach's alpha was computed to check the internal consistency reliability of both the scales. The computed values were 0.78 and 0.86 for community and individuals respectively, revealed acceptable correlation between the items and the corresponding sections of the questionnaire. Exploratory factor analysis was carried out to establish the relation between different variable and group the same. Confirmatory factor analysis was performed to verify the established factors.

Exploratory Factor Analysis (EFA)

The KMO sampling adequacy on the tools were 0.724 for community and 0.78 for individuals, which indicated the sampling sizes were appropriate for performing factor analysis for the tools. Principle Component Analysis (PCA) for the tools produced a 6-factor solution each at individual and community levels which is shown in Table 4.2.

Confirmatory Factor Analysis (CFA)

From CFA, it is evident that the data provided did not differ significantly from the six factor model of EFA ($\chi 2(5) = 10.53$, $p > 0.05$ (NS), RMR = 0.03, GFI = 0.99, AGFI = 0.95, RMSEA = 0.05, NFI = 0.97, CFI = 0.97, and ($\chi 2/df$) = 2.11). All these values fall under the acceptable ranges for corresponding tests. The construct reliability check obtained Composite Reliability score, CR>0.7 which is acceptable. The construct validity checks also yielded desirable results. Convergent validity and Divergent validity were established by computing Average Variance Extracted (AVE), Maximum Shared Variance and Average Shared Variance. AVE values obtained were greater than 0.5 and MSV values. The factor correlations obtained, when compared were less than the square root of AVE values. These results support the conceptualization of a six-factor model for computing ERI.

Table 4.2 EFA split-up of different items under six factors for ERI tool

Sl No.	Factors for ERI for individuals	Item	Variance	Factors for ERI for community	Item	Variance
1	Emergency needs	×12, ×14, ×1, ×20, ×22, ×25	23.96	Community Participation	y1, y2, y3, y4, y5	20.1
2	Personal safety	×9, ×10, ×11, ×16, ×24	12.3	Trust	y13, y14, y15, y16, y17	12.38
3	Outdoor safety	×21, ×26, ×27, ×23	7.24	Collective Efficacy	y8, y9, y10	10.99
4	Indoor safety	×4, ×5, ×7, ×8, ×13	4.86	Willingness	y18, y19, y20	7.98
5	Structural safety	×1, ×2, ×3, ×6	4.33	Empowerment	y11, y12	5.54
6	Mitigation	×15, ×18, ×19	3.91	Intension	y6, y7	5.37

4.5 Classification of Preparedness

Preparedness levels of individuals and community are classified into five levels, namely 'very poor', 'poor', 'moderate', 'good' and 'excellent' based on the ERI scores. Table 4.3 shows the boundary scores based on the percentage of preparedness activities undertaken or completed by the respondents along with the percentage distribution of respondents in each score range. It also gives the degree of preparedness for each levels. The entire continuum was categorized into five intervals. This categorization was adopted for various reasons; it produced a compact set of ranges, the scale generated is unbiased, it ensured sufficient sample sizes in each interval and it made the interpretation simpler. From the given data, it is evident that the degree of preparedness ranges from very poor to moderate for more than 90% respondents at individual level and around 80% at community level.

The ERI intervals with percentage distribution of respondents for ERI for Individuals and ERI for Community are graphically represented in Fig 4.1. Peaks denoting the maximum number of positive responses fall in level 2 for Individuals and level 3 for Community respectively. From the shape of the curves (bell curves) which resemble that of normal distribution, it is clear that the samples are random and are representative of the actual population. The mean scores for ERI for individuals and community are around 10, which falls in level two, indicating poor degree of preparedness. The distributions have scores 11 and 10 against the highest percentage of respondents at individual and community levels with standard deviations of 3.08 and 3.45 respectively.

Table 4.3 Percentage distribution of respondents for various degrees of preparedness with respect to Earthquake Readiness Indices for Individuals and Community

Level	% of preparedness activities done	Degree of preparedness	ERI range for Individuals	% of respondents for ERI for Individuals	ERI range for community	% of respondents for ERI for Community
1	Upto 20	Very poor	0–5	18.89	0–4	9.5
2	Upto 40	Poor	6–10	41.30	5–8	22.5
3	Upto 65	Moderate	11–17	31.11	9–13	47.25
4	Upto 85	Good	18–23	6.48	14–17	16.5
5	Above 85	Excellent	24–27	2.22	18–20	4.25

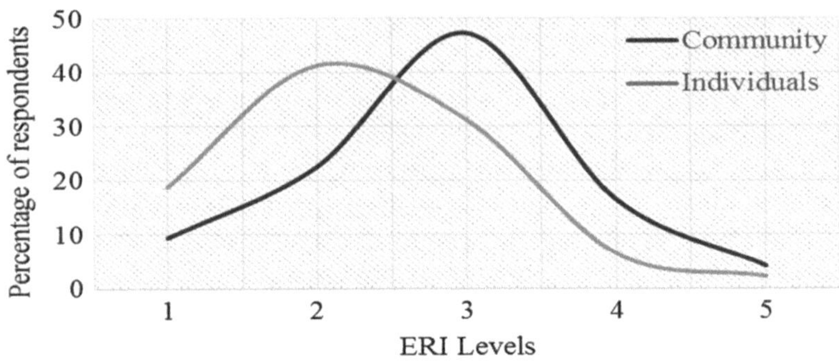

Fig. 4.1 ERI levels with respect to percentage of respondents

Demographics and Preparedness

Analysis and comparison of results based on demographic factors helped to arrive at various conclusions. From the obtained data, it is clear that female population scored better compared to male population at individual level and vice versa in the community level. ERI of individuals increased with age while that of community showed increased indices for age groups 21–25 years and 18–20 years as compared to the rest. The employed group and house persons had better ERI compared to students and unemployed category at individual level, whereas a peak is obtained for the student category at community level. Married persons had obtained higher individual ERI as compared to the rest, while the single category peaks by a high value of 16.7 belonging to level 4 at community level.

It was also observed that ERI increased with the increase in level of educational qualifications. Homeowners had better ERI as compared to non-homeowners. 278 households out of 540 have at least one elderly person or a child at home. Among these only 14.93% have undertaken the emergency needs that are vital in particular to the households with old-age people and children. In this regard more than 85% of the households are not prepared for a major earthquake. Fig 4.2 gives the mean indices based on demographic factors at individual and community levels.

4.6 Conclusion

The ERI tools are designed to accommodate a wide range of earthquake preparedness indicators. The Earthquake Readiness Indices (ERI) are highly valid and reliable tools that assess individuals and community on a six level scale. The indices

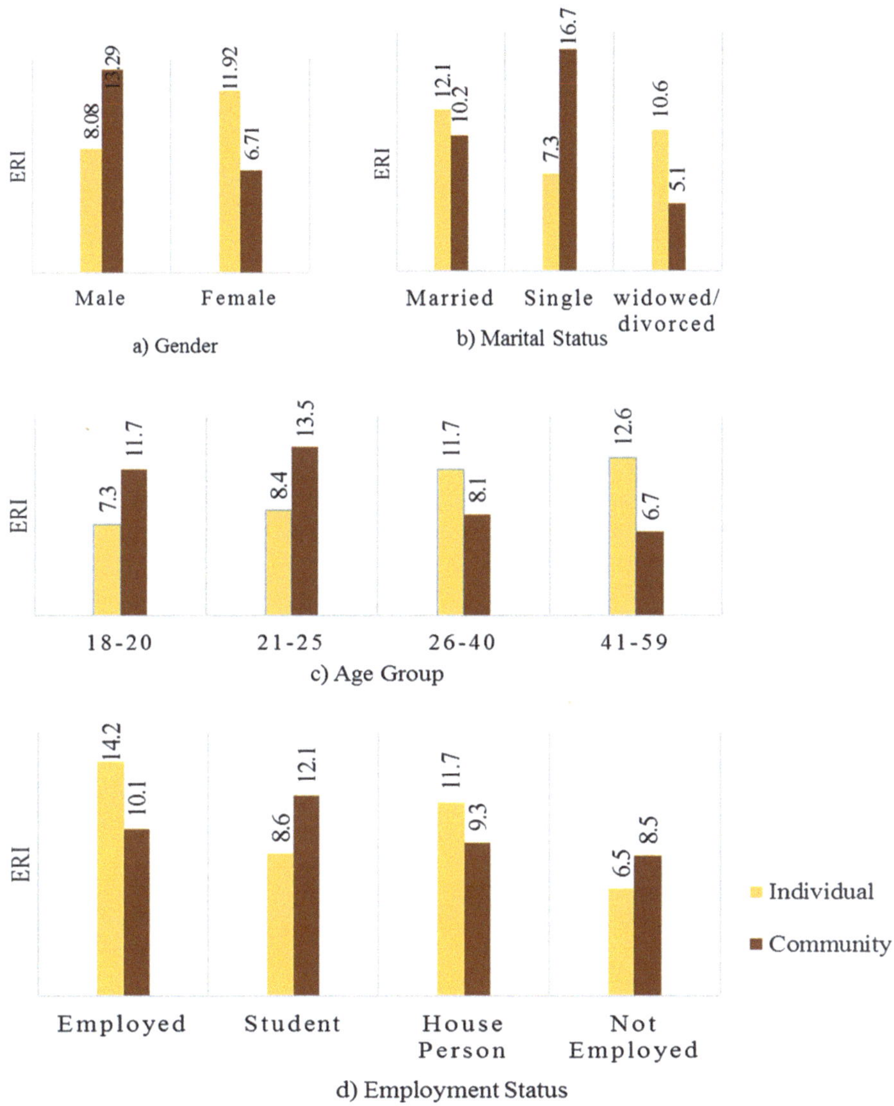

Fig. 4.2 Comparison of mean indices at individual and community level based on demographics

can be expressed in terms of score intervals or as continuous variables ranging from 0-27 and 0-20 for individual and community levels respectively. The former would be practically useful when the interest is on descriptive information. The ERI tool can assist individuals to quantitatively analyse their current preparedness levels. The proposed tool assisted to give a better idea on the levels of preparedness and to identify the areas where people lack awareness. The tool is designed in such a way that the user can identify on which aspects of preparation more effort has to be taken to be actively prepared for a major earthquake.

Appendix: Questionnaire for Earthquake Readiness Indices

Earthquake Readiness Index for Individuals

The following are the things that can be done to minimize the damage and disruption if an earthquake occurs. Record your responses (one response per statement)

How to calculate your score at individual level:

i. Mark your responses against each statement.
ii. Count the total number of responses against the first option (Have done this) for statements 1–25 or 27 (depending on employment status). This gives your ERI score.
iii. The table provided at the end of the questionnaire gives your preparedness category.

1. I have looked into the chance of an earthquake when I decided to live in my present residence.

☐ Have done this ☐ May do this ☐ Will not do this

2. I have ensured tall furniture are fixed against walls.

☐ Have done this ☐ May do this ☐ Will not do this

3. I have ensured heavy utility hardware will not fall over in a major earthquake. (eg: water heaters, chimney, fan, AC).

☐ Have done this ☐ May do this ☐ Will not do this

4. I have made sure that roof will have less chance to collapse in an earthquake.

☐ Have done this ☐ May do this ☐ Will not do this

5. I have organized the cupboards in way that heavy items rest at ground level.

☐ Have done this ☐ May do this ☐ Will not do this

6. I have ensured cupboards are fastened with latches.

☐ Have done this ☐ May do this ☐ Will not do this

7. I have ensured that water containing objects are not stored above 1 meter height. (eg. pot plants or aquariums on top of TV).

☐ Have done this ☐ May do this ☐ Will not do this

8. I have made sure that heavy objects rest on floor.

☐ Have done this ☐ May do this ☐ Will not do this

9. I have made sure that I know the locations of utility switches and shut-off valves (eg. electric power shut off switch, gas shut-off valve, water shut-off valve).

☐ Have done this ☐ May do this ☐ Will not do this

10. I have operated utility switches and shut-off valves.

☐ Have done this ☐ May do this ☐ Will not do this

11. I have stored adequate water for survival for the next 72 hours.

☐ Have done this ☐ May do this ☐ Will not do this

12. I have toilet paper and spare plastic bags for use as an emergency toilet.

☐ Have done this ☐ May do this ☐ Will not do this

13. I have sufficient tools to work on minor repairs in my present house.

☐ Have done this ☐ May do this ☐ Will not do this

14. I have acquired dehydrated or tinned food supply that can be relied on, in an emergency.
☐ Have done this ☐ May d o this ☐ Will not do this

15. I have a first aid box handy.
☐ Have done this ☐ May do this ☐ Will not do this

16. I have kept essential medicines for allergies or illness.
☐ Have done this ☐ May do this ☐ Will not do this

17. I have acquired a functioning battery-operated radio.
☐ Have done this ☐ May do this ☐ Will not do this

18. I have mounted movable objects like television, computer etc on to wall/ desk.
☐ Have done this ☐ May do this ☐ Will not do this

19. I have access to alternate cooking source.
☐ Have done this ☐ May do this ☐ Will not do this

20. I have a household earthquake emergency plan.
☐ Have done this ☐ May do this ☐ Will not do this

21. I have contacts of emergency services (eg: medical, police, support groups)
☐ Have done this ☐ May do this ☐ Will not do this

22. I have acquired a working fire extinguisher.
☐ Have done this ☐ May do this ☐ Will not do this

23. I have acquired alternative charging devices/spare batteries for appliances that are needy. (eg. Torch, mobile phones etc.)
☐ Have done this ☐ May do this ☐ Will not do this

24. I have an emergency kit handy.
☐ Have done this ☐ May do this ☐ Will not do this

25. I have made sure that I can reach out to a nearby medical emergency centre. I have considered the chance of an earthquake when I decided to live in my present house.
☐ Have done this ☐ May do this ☐ Will not do this

If employed/studying, mark the following,

26. I have an earthquake emergency plan at work place/institute.

☐ Have done this ☐ May do this ☐ Will not do this

27. I have taken some other preparedness steps at work place/institute. (Knowing emergency exits, where to be in case of an earthquake).

☐ Have done this ☐ May do this ☐ Will not do this

Earthquake Readiness Index for Community

Note: Community refers to a group in which you belong. It could be your neighbourhood, locality or social groups.

A. **With regard to your life in the community, record your responses. (Questions 1-5). (One response per statement)**

How to calculate your score at community level:

i. Mark your responses against each statement.
ii. Count the total number of responses against the 1st option (Have done this) for statements 1 to 10.
iii. Count the total number of responses against 1st (5) and 2nd (4) options for statements 11-20.
iv. Sum of ii and iii gives the ERI score at community level.
v. The table provided at the end of the questionnaire gives your preparedness category.

1. I have worked jointly with other for better community life.

☐ Have done this ☐ May do this ☐ Will not do this

2. I have participated in local events or activities (eg. fairs, festivals).

☐ Have done this ☐ May do this ☐ Will not do this

3. I have given money to charities or social causes in my community.

☐ Have done this ☐ May do this ☐ Will not do this

4. I have participated in public meetings regarding community issues.

☐ Have done this ☐ May do this ☐ Will not do this

5. I have taken part in volunteer activities beneficial to my community (eg. community groups, fund-raising, Scouts, clean-up days). I have worked with others to improve community life.

☐ Have done this ☐ May do this ☐ Will not do this

B. With regard to preparedness point of view, record your responses. (Questions 6-10). (One response per statement)

6. Participate on an on-going basis regularly. (eg., attend monthly meeting, belong to a group).

☐ Have done this ☐ May do this ☐ Will not do this

7. Participate for specific events or reasons. (eg. attend community meetings).

☐ Have done this ☐ May do this ☐ Will not do this

8. Pass on information regarding earthquake and preparedness to other members of the community.

☐ Have done this ☐ May do this ☐ Will not do this

9. Encourage other community members to get prepared for earthquakes?

☐ Have done this ☐ May do this ☐ Will not do this

10. Take training to be qualified to assist your community recover from the after effects of earthquakes. Participate regularly, on an on-going basis (e.g. belong to a group; attend monthly meetings)?

☐ Have done this ☐ May do this ☐ Will not do this

C. **In regard to your general feelings about living in this community, please record to which extent you agree or disagree with the following statements on a scale of 1-5** ***(5- Strongly agree, 4- agree, 3- neither agree nor disagree, 2- disagree , 1- Strongly disagree)*** **(Questions 11-20).(one response per statement)**

11.	I am often a leader in community meeting.	☐5	☐4	☐3	☐2	☐1
12.	I prefer leadership in community meetings.	☐5	☐4	☐3	☐2	☐1
13.	In my community, people will express opinion for relevant issues.	☐5	☐4	☐3	☐2	☐1
14.	In my community, people can express their needs to others easily.	☐5	☐4	☐3	☐2	☐1
15.	The talks of community depends on residents' interests.	☐5	☐4	☐3	☐2	☐1
16.	The focus of the community depends on difficult situations.	☐5	☐4	☐3	☐2	☐1
17.	People's outlook of community problems control the measures adopted.	☐5	☐4	☐3	☐2	☐1
18.	I trust the local government to provide timely and appropriate response to the necessities of its people.	☐5	☐4	☐3	☐2	☐1
19.	I trust the mass media to broadcast and report reliable information.	☐5	☐4	☐3	☐2	☐1
20.	I trust the law to maintain order and take care of my community.	☐5	☐4	☐3	☐2	☐1

Demographics

We assure you that all information that you provide us will be strictly confidential and it will be restricted to the purpose of the study.

1. Your gender: ☐ Male ☐ Female ☐ Prefer not to say ☐ Other

2. Which category below includes your age?
   ☐ 17 or younger ☐ 18-20 years ☐ 21-25years
   ☐ 26-40 years ☐ 41-59 years ☐ 60 years or older

3. What is your highest educational qualification?
   ☐ Less than high school ☐ High school ☐ Higher Secondary
   ☐ Bachelor's degree ☐ Master's degree ☐ Doctoral degree
   ☐ Other (please specify) ___________

4. Which of the following category best describes your employment status?
   ☐ Employed (full-time) ☐ Employed (part-time) ☐ Self-employed Student ☐ House person ☐ Retired ☐ Not employed ☐ Disabled, not able to work

5. What is your current household annual income?
   ☐ Less than ₹ 27,000 ☐ Above ₹ 27,000 to ₹ 1 lakh
   ☐ Above ₹ 1 lakh to ₹ 3 lakhs ☐ Above ₹ 3 lakhs to ₹ 7 lakhs
   ☐ Above ₹ 7 lakhs to ₹ 25 lakhs ☐ Above ₹ 25 lakhs to ₹ 1 crore
   ☐ Above ₹ 1 crore

6. How long have you lived in your current residence? (in years) _____

7. Including yourself , how many people live in your current residence?_____

8. Which of the following describes your family status?
   ☐ Married ☐ Single ☐ Divorced
   ☐ Separated ☐ Widowed ☐ Would rather not say

9. How many members above 60 year old live in your current residence?___

10. How many children under 16 year old live in your current residence?_____

11. Which of the following describes the ownership of the house you live in?
    ☐ Owned or buying to live in it ☐ Owned or buying as a holiday home
    ☐ Rent to live in it ☐ Rent, as a holiday home

12. Which is your native city? _________

Preparedness levels

Level	ERI range for individuals	ERI range for community	Degree of preparedness	% of preparedness activities done
1	0–5	0–4	Very poor	Upto 20
2	6–10	5–8	Poor	Upto 40
3	11–17	9–13	Moderate	Upto 65
4	18–23	14–17	Good	Upto 85
5	24–27	18–20	Excellent	Above 85

Institutes/Organizations in India Involved in Earthquake Studies/Management

- Central Building Research Institute, www.cbri.res.in/
- Department of Earthquake Engineering, IIT Roorkee http://www.iitr.ac.in/departments/EQ/pages/
- Earthquake Engineering Research Center, IIIT Hyderabad, http://eerc.iiit.ac.in/
- Geological Survey of India http://www.portal.gsi.gov.in/
- Indian Meteorological Department (IMD) www.imd.gov.in/
- Institute of Seismological Research (ISR), http://www.isr.gujarat.gov.in/
- Ministry of Earth Sciences, Govt. of India www.moes.gov.in/
- National Disaster Management Authority (NDMA) www.ndma.gov.in/
- National Disaster Response Force (NDRF) http://ndrfandcd.gov.in/cms/Ndrf.aspx
- National Geophysical Research Institute www.ngri.org.in/
- National Information Centre of Earthquake Engineering (NICEE), IIT Kanpur www.nicee.org/
- National Institute of Disaster Management (NIDM) http://nidm.gov.in/
- Regional Meteorological Centre, Mumbai: Earthquakes http://www.imdmumbai.gov.in/

Registered Societies in India

- Indian Society Of Earthquake Science (http://www.ises.in/)
- ISET: Indian Society of Earthquake Technology (http://iset.org.in/)

T.G. Sitharam, S. Kolathayar, *Preparing for Earthquakes: Lessons for India*, SpringerBriefs in Environmental Science, DOI 10.1007/978-3-319-59522-1

References

Abhishek, K., Anbazhagan, P., & Sitharam, T. G. (2013). Seismic hazard analysis of Lucknow considering local and active seismic gaps. *Natural Hazards, 69*, 327–350.

Algermissen, S. T., & Perkins, D. M. (1976). *A probabilistic estimate of maximum acceleration in rock in the contiguous United States, U.S. Geological Survey*. (Open-file Report, 76–416), 45 pp.

Ardalan, A., & Sohrabizadeh, S. (2016). Assessing households preparedness for earthquakes: An exploratory study in the development of a valid and reliable Persian-Version Tool. *PLOS Current*. doi:10.1371/currents.dis.ccc8697279713e66887b928b839d0920.

Baytiyeh, H. (2014). *The assessment of earthquake preparatory knowledge and activities of Lebanese engineering students*. Proceedings of American Society for Engineering Education Annual Conference and Exposition, 14, 11978–11987.

Becker, J. S. (2010) *Understanding disaster preparedness and resilience in Canterbury: Results of interviews, focus groups and a questionnaire survey*. GNS Science Report.

Becker, J. S., Paton, D., Johnston, M. D., & Ronan, K. R. (2012). A model of household preparedness for earthquakes: How individuals make meaning of earthquake information and how this influences preparedness. *Natural Hazards, 64*, 107–137.

Bhatia, S. C., Kumar, M. R., & Gupta, H. K. (1999). A probabilistic seismic hazard map of India and adjoining regions. *Annali di Geofisica, 42*, 1153–1166.

Bilham, R. (2004). Earthquakes in India and the Himalaya: Tectonics, geodesy and history. *Annals of Geophysics, 47*, 839–858.

Boominathan, A., Dodagoudar, G. R., Suganthi, A., & Uma Maheswari, R. (2008). Seismic hazard assessment of Chennai city considering local site effects. *Journal of Earth System Science, 117*(S2), 853–863.

Chandra, U. (1977). Earthquakes of Peninsula India—A seismotectonic study. *Bulletin of the Seismological Society of America, 67*, 1387–1413.

Curray, J. R. (2005). Tectonics and history of the Andaman Sea region. *Asian Journal of Earth Sciences, 25*, 187–228.

Dasgupta, S., Mukhopadhyay, B., & Acharyya, A. (2007). Seismotectonic of Andaman–Nicobar region: Constraints from aftershocks within 24 hours of the great 26 December 2004 earthquake. *Geological Survey of India Special Publication, 89*, 95–104.

Dasgupta, S., Mukhopadhyay, B., & Bhattacharya, A. (2007). Seismicity pattern in north Sumatra—Great Nicobar region: In search of precursor for the 26 December 2004 earthquake. *Journal of Earth System Science, 116*(3), 215–223.

Dasgupta, S., & Mukhopadhyay, M. (1993). Seismicity and plate deformation below the Andaman Arc, Northeast Indian Ocean. *Tectonophysics, 225*, 529–542.

T.G. Sitharam, S. Kolathayar, *Preparing for Earthquakes: Lessons for India*, SpringerBriefs in Environmental Science, DOI 10.1007/978-3-319-59522-1

Dasgupta, S., & Mukhopadhyay, M. (1997). Aseismicity of the Andaman subduction zone and recent volcanism. *Journal of the Geological Society of India, 49*, 513–521.
Dasgupta, S., Mukhopadhyay, M., Bhattacharya, A., & Jana, T. K. (2003). The geometry of the Burmese–Andaman subducting lithosphere. *Journal of Seismology, 7*, 155–174.
De Man, A., & Simpson-Housley, P. (1987). Factors in the perception of earthquake hazard. *Perceptual and Motor Skills, 64*, 815–820.
Deshikachar, S. V. (1974). A review of the tectonic and geological history of eastern India in terms of "plate tectonics" theory. *Journal of the Geological Society of India, 15*, 137–149.
Dooley, D., Catalano, R., Mishra, S., & Serxner, S. (1992). Earthquake preparedness: Predictors in a community survey. *Journal of Applied Social Psychology, 22*, 451–470.
Edwards, M. E. (1993). Social location and self-protective behavior: Implications for earthquake preparedness. *International Journal of Mass Emergencies Disasters, 11*, 293–303.
Eguchi, T., Uyeda, S., & Maki, T. (1979). Seismotectonics and tectonic history of Andaman sea. *Tectonophysics, 57*, 35–51.
Evans, P. (1964). The tectonic framework of Assam. *Journal of the Geological Society of India, 5*, 80–96.
Farley, J. E., Barlow, H. D., Finklestein, M. S., & Riley, L. (1993). Earthquake hysteria, before and after: A survey and follow-up on public response to the Browning forecast. *International Journal of Mass Emergencies Disasters, 11*, 305–322.
Gangrade, B. K., & Arora, S. K. (2000). Seismicity of the Indian Peninsular Shield from regional earthquake data. *Pure and Applied Geophysics, 157*, 1683–1705.
Giardini, D. (1984). Systematic analysis of deep seismicity—200 centroid-moment tensor solutions for earthquakes between 1977 and 1980. *Geophysical Journal of the Royal Astronomical Society, 77*, 883–914.
Giardini, D., Grünthal, G., Shedlock, K. M., & Zhang, P. (1999). The global seismic hazard assessment program (GSHAP). *Annali di Geofisica, 42*, 1225–1228.
Gupta, I. D. (2006). Delineation of probable seismic sources in India and neighbourhood by a comprehensive analysis of seismotectonic characteristics of the region. *Soil Dynamics and Earthquake Engineering, 26*, 766–790.
Hirose, H. (1986). The psychological impact of the Tokai earthquake prediction: Individual's responses and the mass media coverage. *Japanese Psychological Research, 26*, 64–76.
Hirose, H., & Ishizuka, T. (1983). Causal analysis of earthquake concern and preparing behavior in the North Izu Peninsula. *Japanese Psychological Research, 25*, 103–111.
Hurnen, F., & McClure, J. (1997). The effect of increased earthquake knowledge on perceived preventability of earthquake damage. *Australasian Journal of Disaster and Trauma Studies, 3*, 1–14.
IS-1893. (2002). *Indian standard criteria for earthquake resistant design of structures, Part 1—General provisions and buildings*. New Delhi: Bureau of Indian Standards Bureau of Indian Standards.
Iyengar, R. N., & Ghosh, S. (2004). Microzonation of earthquake hazard in greater Delhi area. *Current Science, 87*, 1193–1202.
Iyengar, R. N., et al. (2010). *Development of probabilistic seismic hazard map of India*. Technical report of the Working Committee of Experts (WCE) constituted by the National Disaster Management Authority. New Delhi: Govt. of India.
Jackson, E. L. (1981). Response to earthquake hazard: The West Coast of North America, *Environmental. Behaviour, 14*, 387–416.
Jackson, E. L., & Mukerjee, T. (1974). *Human adjustment to the earthquake hazard in San Francisco, California*. In G. F. White (Ed.), *Natural hazards: Local national and global* (pp. 160–166). New York: Oxford University Press.
Jaiswal, K., & Sinha, R. (2007). Probabilistic seismic-hazard estimation for peninsular India. *Bulletin of the Seismological Society of America, 97*(1B), 318–330.
Joyner, W. B., & Boore, D. M. (1981). Peak horizontal acceleration and velocity from strong-motion records including records from the 1979 Imperial Valley, California, earthquake. *Bulletin of the Seismological Society of America, 71*, 2011–2038.

Kayal, J. R. (2008). *Microearthquake seismology and seismotectonics of South Asia* (1st ed.). New Delhi: Springer. ISBN:978-1-4020-8179-8.

Khattri, K. N., Rogers, A. M., Perkins, D. M., & Algermissen, S. T. (1984). A seismic hazard map of India and adjacent areas. *Tectonophysics, 108*, 93–134.

Kiecolt, K. J., & Nigg, J. M. (1982). Mobility and perceptions of a hazardous environment. *Environmental Behaviour, 14*, 131–154.

Kolathayar, S., & Sitharam, T. G. (2012a). Characterization of regional seismic source zones in and around India. *Seismological Research Letters, 83*(1), 77–85.

Kolathayar, S., & Sitharam, T. G. (2012b). Comprehensive probabilistic seismic hazard analysis of Andaman-Nicobar regions. *Bulletin of Seismological Society of America (SSA), 102*(5), 2063–2076.

Kolathayar, S., Sitharam, T. G., & Vipin, K. S. (2011). Spatial variation of seismicity parameters across India and adjoining areas. *Natural Hazards, 60*(3), 1365–1379.

Kolathayar, S., Sitharam, T. G., & Vipin, K. S. (2012). Deterministic seismic hazard macrozonation of India. *Journal of Earth System Sciences, 121*(5), 1351–1364.

Krishnan, M. S. (1953). The structure and tectonic history of India. *Memoirs—Geological Survey of India, 81*, 137.

Kumar, P., Yuan, X., Ravi Kumar, M., Kind, R., Li, X., & Chadha, R. K. (2007). The rapid drift of Indian tectonic plate. *Nature, 449*, 894–897.

McClure, J., & Allen, W. (2001). When earthquake damage is seen as preventable: Attributions, locus of control and attitudes to risk. *Basic and Applied Social Psychology, 23*(2), 109–121.

McClure, J., Fischer, R., Hunt, M., & Charleson, A. (2007). *Using action plans to increase voluntary actions that reduce earthquake damage*. EQC Report No. 2314, Earthquake Commission, Wellington, New Zealand.

McClure, J., Sibley, C. G., & White, J. (2009). Framing effects on preparation intentions: Distinguishing actions and outcomes. *Disaster Prevention and Management, 18*(2), 187–199.

Menon, A., Ornthammarath, T., Corigliano, M., & Lai, C. G. (2010). Probabilistic seismic hazard macrozonation of Tamil Nadu in Southern India. *Bulletin of the Seismological Society of America, 100*, 1320–1341.

Mileti, D., & O'Brien, P. W. (1992). Warnings during disasters: Normalizing communicating risk'. *Social Problems, 38*, 40–57.

Mohanty, W. K., & Walling, M. Y. (2008). First order seismic microzonation of Haldia, Bengal basin (India) using a GIS platform. *Pure and Applied Geophysics, 165*(7), 1325–1350.

Molnar, P., & Tapponnier, P. (1979). The collision between India and Eurasia. In B. A. Bolt (Ed.), *Earthquakes and volcanoes, Procs., from Scientific American* (pp. 62–73). San Francisco: WH Freeman.

Mukhopadhyay, M. (1984). Seismotectonics of subduction and back-arc rifting under the Andaman Sea. *Tectonophysics, 108*, 229–239.

Mukhopadhyay, M. (1988). Gravity anomalies and deep structure of Andaman arc. *Marine Geophysical Research, 9*, 197–210.

Mulilis, J. P., Duval, T. S., & Bovalino, K. (1990). Tornado preparedness of students, nonstudent renters, and non-student owners: Issues of PrE theory. *Journal of Applied Social Psychology, 30*, 1310–1329.

Mulilis, J. P., & Lippa, R. (1990). Behavioral change in earthquake preparedness due to negative threat appeals: A test of protection motivation theory. *Journal of Applied Social Psychology, 20*, 619–638.

Narayan, J. P., Sharma, M. L., & Kumar, A. (*2002*). A seismological report on the January 26, 2001 Bhuj, India earthquake. *Seismological Research Letters, 73*(3), 343–355.

Nath, S. K., & Thingbaijam, K. K. S. (2010). Peak ground motion predictions in India: An appraisal for rock sites. *Journal of Seismology, 15*(2), 295–315.

Nath, S. K., & Thingbaijam, K. K. S. (2012). Probabilistic seismic hazard assessment of India. *Seismological Research Letters, 83*(1), 135–149.

Okazaki, K., Ilki, A., Ahmad, N., Kandel, R.C., & Rahayu, H. (2008). *Seismic risk perception of people for safer housing*. Proceedings of the 14th World Conference on Earthquake Engineering.

Parvez, I. A., Vaccari, F., & Panza, G. F. (2003). A deterministic seismic hazard map of India and adjacent areas. *Geophysics Journal International, 155*, 489–508.
Paton, D., Bajek, R., Okada, N., & McIvor, D. (2010). Predicting community earthquake preparedness: A cross-cultural comparison of Japan and New Zealand. *Natural Hazards, 54*, 765–781.
Purnachandra Rao, N. (1999). Single station moment tensor inversion for focal mechanisms of Indian intra-plate earthquakes. *Current Science, 77*, 1184–1189.
Raghu Kanth, S. T. G., & Iyengar, R. N. (2006). Seismic hazard estimation for Mumbai city. *Current Science, 91*(11), 1486–1494.
Rai, D. C., & Murty, C. V. R. (2003). *Reconnaissance report: North Andaman (Diglipur) earthquake of 14 September 2002*. New Delhi: Dept. of Science and Technology, Government of India, New Delhi, United Nations Development Programme.
Rajendran, K., & Gupta, H. K. (1989). Seismicity and tectonic stress field of a part of the Burma-Andaman-Nicobar Arc. *Bulletin of the Seismological Society of America, 79*, 989–1005.
Rao, B. V., & Murty, B. V. S. (1970). Earthquakes and tectonics in peninsular India. *Journal of Indian Geophysical Union, 7*, 1–8.
Reddy, P. R. (2003). Need for high-resolution deep seismic reflection studies in strategic locales of South India. *Current Science, 84*(8), 973–974.
Russell, L. A., & Bourque, L. B. (1995). Preparedness and hazard mitigation actions before and after two earthquakes. *Environment and Behavior, 27*, 744–770.
Scholz, C. H. (1988). Mechanisms of seismic quiescences. *Pure and Applied Geophysics, 126*, 701–718.
Seeber, L., Armbruster, J. G., & Jacob, K. H. (1999). *Probabilistic assessment of earthquake hazard for the state of Maharashtra.*. Report to Government of Maharashtra Earthquake Rehabilitation Cell, Mumbai.
SEISAT. (2000). *Seismotectonic atlas of India and its Environs*. Calcutta: *Geological Survey of India*.
Sharma, M. L. (*2003*). Seismic hazard in northern India region. *Seismological Research Letters, 74*(2), 140–146.
Sharma, M. L., & Malik, S. (2006). *Probabilistic seismic hazard analysis and estimation of spectral strong ground motion on Bed rock in North East India*. 4th International Conference on Earthquake Engineering, Taipei, Taiwan, Oct 12–13, 2006.
Sil, A., Sitharam, T. G., & Kolathayar, S. (2013). Probabilistic seismic hazard analysis of Tripura and Mizoram states. *Natural Hazards, 68*(2), 1089–1108.
Singh, R. K. (2009). *Probabilistic seismic hazard and risk analysis: A case study for Ahmedabad city*. Ph.D. thesis, Indian Institute of Technology, Kanpur.
Sitharam, T. G., & Anbazhagan, P. (2007). Seismic hazard analysis for the Bangalore region. *Natural Hazards, 40*, 261–278.
Sitharam, T. G., & Kolathayar, S. (2013). Seismic hazard analysis of India using areal sources. *Journal of Asian Earth Sciences, 62*, 647–653.
Sitharam, T. G., Kolathayar, S., & James, N. (2015). Probabilistic assessment of surface level seismic hazard in India using topographic gradient as a proxy for site condition. *Geoscience Frontiers, 6*(6), 847–859.
Spittal, M. J., Mc Clure, J., Siegert, R. J., & Walkey, F. H. (2008). Predictors of two types of earthquake preparation: Survival activities and mitigation activities. *Environment and Behavior, 40*(6), 798–817.
Spittal, M. J., Mc Clure, J., Siegert, R. J., Walkey, F. H., & Ballantyne, K. E. (2006). The earthquake readiness scale: The development of a valid and reliable unifactorial measure. *Natural Hazards, 39*, 15–29.
Sullivan, R., Mustart, D. A., & Galehouse, J. S. (1977). Living in earthquake country. *California Geology, 3*, 3–8.
TC4-ISSMGE. (1999). *Manual for zonation on seismic geotechnical hazard* (Revised edition). Technical Committee for Earthquake Geotechnical Engineering (TC4) of the International Society of Soil Mechanics and Geotechnical Engineering (ISSMGE) 209.

Turner, R. H., Nigg, J. M., & Paz, D. H. (1986). *Waiting for disaster: Earthquake watch in California*. Los Angeles: University of California Press.

Vipin, K. S., Anbazhagan, P., & Sitharam, T. G. (2009). Estimation of peak ground acceleration and spectral acceleration for South India with local site effects: Probabilistic approach. *Natural Hazards and Earth System Sciences, 9*, 865–878.

Yang S., Gao H., Liu, L., Tang, W., He, L., & Fan, C. (2010). *Analysis on public earthquake risk perception: Based on questionnaire*. Proceedings of 3rd International Conference on Cartography and GIS, Nessebar, Bulgaria.

About the Authors

T.G. Sitharam, F.I.G.S., F.I.E., completed his masters from Indian Institute of Science (in 1986) and Ph.D. from the University of Waterloo, Waterloo, ON, Canada (in 1991). He is KSIIDC Chair Professor in the area of Energy and Mechanical Sciences and Senior Professor in the Department of Civil Engineering, Indian Institute of Science, Bengaluru (IISc). He was former founding Chairman of the Center for Infrastructure, Sustainable Transport and Urban Planning (CiSTUP) at IISc. He is presently the Chairman of the AICTE South Western Zonal Committee, regional office at Bengaluru, and Vice President of the Indian Society for Earthquake Technology (ISET). He was a Visiting Professor at Yamaguchi University, Japan, and ISM Dhanbad, Jharkhand. Prof. Sitharam has served as convener for developing Geotechnical guidelines for seismic microzonation supported by National Disaster Management Authority (NDMA), Govt. of India, New Delhi. His work has been recognized by Indian Geotechnical Society (IGS) through IGS-Kucklemann award in 2015 and by IIT Roorkee through Prof. Gopal Ranjan research award in 2014 for his lifetime contributions in Geotechnical Engineering.

Over the last 25 years, he has carried out seismic microzonation of urban centers in India and developed innovative technologies in the area of fracturing and geotechnical applications, leading to about 500 technical papers, 7 books, 3 patents, 100 consulting projects, and 2 startup companies. He is the chief editor of two international journals in his areas of research and founder President of the International Association for Coastal Reservoir Research (IACRR).

Sreevalsa Kolathayar pursued his M.Tech from IIT Kanpur, Ph.D. from Indian Institute of Science (IISc), and served as International Research Staff at UPC BarcelonaTech, Spain, with Marie Curie Fellowship by European Union. He is presently Assistant Professor and Research Coordinator in the Department of Civil Engineering, Amrita Vishwa Vidyapeetham University, Coimbatore, India. He

T.G. Sitharam, S. Kolathayar, *Preparing for Earthquakes: Lessons for India*,
SpringerBriefs in Environmental Science, DOI 10.1007/978-3-319-59522-1

served as Chairman of Students Council at IISc and acts as Adviser to Think India, a national forum of students from premier Institutes in India. Dr. Sreevalsa has published several research articles focusing on earthquake hazard and risk assessment. His team has developed a survey-based tool and mobile application "BhookampRaksha" to measure and boost the Earthquake Readiness Index of the public. In the year 2017, Dr. Sreevalsa was featured by the New Indian Express in their Edex anniversary edition and was honored with Edex award: 40 under 40—South India's Most Inspiring Young Teachers.

Index

T.G. Sitharam, S. Kolathayar, *Preparing for Earthquakes: Lessons for India*, SpringerBriefs in Environmental Science, DOI 10.1007/978-3-319-59522-1

FSC
www.fsc.org
MIX
Papier | Fördert gute Waldnutzung
FSC® C083411

Kolibri 360 GmbH
Ferdinand-Jühlke-Straße 7
99095 Erfurt, Deutschland
produktsicherheit@kolibri360.de